W9-BXD-152

ORGANIC NOMENCLATURE
A PROGRAMMED INTRODUCTION

ORGANIC NOMENCLATURE
A PROGRAMMED INTRODUCTION

FIFTH EDITION

JAMES G. TRAYNHAM
Professor of Chemistry, Emeritus
Louisiana State University

Prentice Hall
Upper Saddle River, New Jersey 07458

Library of Congress Cataloging-in-Publication Data

Traynham, James G.
   Organic nomenclature: a programmed introduction / James G. Traynham—5th ed.
      p. cm.
   Includes bibliographical references.
   ISBN 0–13–270752–7
   1. Chemistry, Organic—Nomenclature—Programmed instruction. I. Title.
QD291.T72   1997
547'.001'4—dc21   96–45238
                                                                                                   CIP

ACQUISITION EDITOR: *John Challice*
PRODUCTION EDITOR: *Joanne E. Jimenez*
MANUFACTURING MANAGER: *Trudy Pisciotti*
PAGE AND ART COMPOSITION: *Eric Hulsizer*
COVER DESIGNER: *Bruce Kenselaar*

© 1997, 1991, 1985, 1980, 1966 by Prentice-Hall, Inc.
Simon & Schuster / A Viacom Company
Upper Saddle River, NJ 07458

All rights reserved. No part of this book may be reproduced, in any form
or by any means, without permission in writing from the publisher.

Printed in the United States of America.

10 9 8 7 6 5 4 3 2

ISBN 0-13-270752-7

Prentice Hall International (UK) Limited, *London*
Prentice Hall of Australia Pty. Limited, *Sydney*
Prentice Hall of Canada, Inc., *Toronto*
Prentice Hall Hispanoamericana, S.A., *Mexico*
Prentice Hall of India Private Limited, *New Delhi*
Prentice Hall of Japan, Inc., *Tokyo*
Simon & Schuster Asia Pte. Limited, *Singapore*
Editora Prentice Hall do Brasil, Ltda., *Rio de Janeiro*

To Professor C.D. Hurd,

and to the memory of the late Professor R. K. Summerbell

—two who were models for combining praise-winning undergraduate teaching and praise-winning research,

and to the IUPAC Commission on Nomenclature of Organic Chemistry,

the members of which are models of devoted service

to the profession of chemistry.

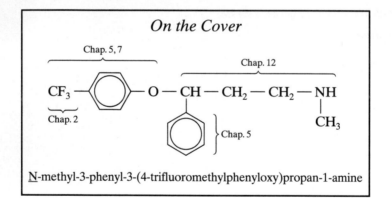

On the Cover

Chap. 5, 7

Chap. 12

Chap. 2

CF₃ — O — CH — CH₂ — CH₂ — NH

Chap. 5

CH₃

N-methyl-3-phenyl-3-(4-trifluoromethylphenyloxy)propan-1-amine

Contents

Preface

The first nomenclature session, as reported in Genesis, must have been easy: Adam had only himself to satisfy, and no rival systems of nomenclature were in use. By contrast, the many sessions concerned with organic chemical nomenclature during the past century, most as activities of the International Union of Pure and Applied Chemistry Commission on Nomenclature of Organic Chemistry (IUPAC CNOC), have had to deal with existing systems, each with its own advocates, with parochial proposals, and with resistance to change from the familiar.

Chemical nomenclature is essential for convenient communication of chemical information. The official IUPAC rules for chemical nomenclature are intended to facilitate unambiguous chemical communication rather than to generate the shortest identifying tag. Registry numbers generated by Chemical Abstracts Service (CAS) are brief, unambiguous identifications for chemical substances, but they are not substitutes for names in everyday use any more than our Social Security numbers are. We continue to need and use names, both for persons and for chemical compounds.

A chemical name identifies the chemical, but it can do much more. The official rules of organic chemical nomenclature have grown out of emphases beyond mere identification, emphases on conveying information primarily about structure and about the expected chemical behavior of the compound named. Extensive investigations of organic chemical behavior, however, had been described in the chemical literature before structure for compounds, as we know it, was even acknowledged, much less known. Names not based on structure were devised and used.

Systematic organic chemical nomenclature has emerged slowly and is still unfinished business. The first official effort was made in 1892, when 34 prominent chemists from nine European countries met in Geneva, Switzerland, as an International Commission for the Reform of Chemical Nomenclature. They adopted a page-and-a-half of 46 rules (resolutions), mainly for aliphatic compounds. Prominent among these rules was the selection of the longest continuous chain of carbon atoms as the basis for a substitutive name and the advocacy of a single name for each compound. These Geneva rules gained substantial, but not complete, acceptance and usage. Some old, familiar names, not reflective of structure, persisted in usage, and new compounds presented problems not considered at the Geneva meeting.

Subsequent official development, refinements, changes, and expansions of the rules have been made through the unpaid work of the CNOC, which has, through several decades, "confined its efforts to codifying sound practices which already existed, rather than originating new nomenclature." Even for new kinds of compounds, only usage that does not conflict with approved rules that are consistent for a large number of compounds is likely to be considered sound practice and be codified. The IUPAC CNOC has been stimulated and guided by proposals from individuals and chemical society committees interested in a particular area of organic chemistry, as well as by the indexing practice of Chemical Abstracts Service (CAS).

The latest comprehensive collection of rules for the nomenclature of organic chemistry is a 559-page book published in 1979 and supplemented by subsequent publications focused on special sections such as radicals and ions or natural products. The latest comprehensive supplement is *A Guide to IUPAC Nomenclature of Organic Compounds*, published in 1993 after years of intense study and debate by CNOC with consideration of comments and suggestions from organic chemists around the world. That *Guide* introduced approved innovations in organic nomenclature that are intended to make naming practice easier and more consistent. The 1979 rules have not been replaced, but new style names are an alternative that may become dominant as they become familiar to chemists. Both the traditional (1979) and the new (1993) style names are included in this book, because both are correct, IUPAC-approved names. Students are in a transition time. Knowledge and understanding of the traditional names are essential for use of past and current literature in chemistry. Knowledge and understanding of the new names are essential for current and future uses. The work toward completely satisfactory nomenclature rules continues and will have to do so as long as new chemical species are being created. Over time chemists get better ideas about nomenclature, just as they do about chemical synthesis and chemical reactivity.

The IUPAC rules are actually more permissive among styles of names than many authors seem to believe. tert-Butyl alcohol and 2-methyl-2-propanol are both IUPAC-approved names for the same compound, the first illustrating radicofunctional nomenclature (class designation—alcohol—a separate word) and the second illustrating substitutive nomenclature (substituents modifying a parent compound, not separated). Several other styles are also approved and used for other kinds of compounds. It is important to realize that while a compound (structure) may be named correctly by more than one name, a correct name always implies only one structure. Some styles are more convenient in one context, others in another context. CAS with its strong emphasis on indexing and retrieval, especially by on-line technology, restricts its index names to one for each compound. More often than not, this name illustrates substitutive nomenclature and IUPAC rules. Sometimes, however, CAS uses an index name that does not follow IUPAC rules exactly. Even then, a person familiar with IUPAC rules for substitutive names is unlikely to find CAS names ambiguous.

Some naming problems will usually require diligent reference to the complete set of rules (just as some chemical transformation problems require diligent reference to the archival literature), but many commonly encountered molecular assemblies can be named unambiguously by conscientious, attentive use of a small set of rules. This book focuses attention on those rules through a format that encourages analysis, pattern recognition, and writing while studying—study skills that pay off in all aspects of an organic chemistry course, not just in learning nomenclature. I'd like to thank all those who provided comments on the previous edition, especially John Langston, Yuba College.

Although changes in this edition have been made to reflect the shift in current IUPAC preferences and to incorporate the 1993 recommendations, the purpose of the book remains unchanged: to help lower the barrier to competence in the use of unambiguous organic chemical nomenclature.

James G. Traynham

ORGANIC NOMENCLATURE
A PROGRAMMED INTRODUCTION

1
Alkanes

Learning nomenclature as well as the chemical behavior of organic compounds is greatly simplified when the compounds are divided into classes. Classification depends on the types of bonds between atoms. Perhaps the simplest class of organic compounds is that of the <u>alkanes</u>, compounds composed solely of carbon and hydrogen with only single bonds between pairs of atoms. <u>Alkanes</u> may also be called <u>hydrocarbons</u>, a name that signals the combination of <u>hydrogen</u> and <u>carbon</u>. <u>Hydrocarbon</u> is a name indicating only the types of atoms present; <u>alkane</u> indicates not only the type of atoms but also the type of bonds that bind them together (only single bonds between pairs of atoms).

In stable organic compounds, the valence, or bonding capacity, of carbon is four, and the valence of hydrogen is one. The simplest alkane contains one carbon and four hydrogens and can be represented by the structural formula

$$
\begin{array}{c}
H \\
| \\
H-C-H \\
| \\
H
\end{array}
$$

An alkane containing two carbons can be represented by the structural formula

$$
\begin{array}{c}
HH \\
|| \\
H-C-C-H \\
|| \\
HH
\end{array}
$$

Note that both of these formulas indicate four bonds to each carbon and one bond to each hydrogen.

1. An alkane containing three carbons can be represented by the structural formula

_____ .

2. This formula indicates _____ bond(s) to each carbon and _____ bond(s) to each hydrogen.
   (number) (number)

3. For convenience in writing, condensed structural formulas are used most frequently. The carbons are still written separately, but hydrogens bound to each carbon are not. Condensed structural formulas for

alkanes containing one, two, and three carbons, respectively, are CH_4, $CH_3 - CH_3$, and _____.

4. Note that the condensed structural formulas still indicate the correct valence for each atom. Counting each hydrogen as one, we find that the number of bonds indicated for each terminal carbon

in $CH_3 — CH_2 — CH_3$ is _____ and for the center carbon is _____.
                     (number)   (number)

Condensed structural formulas rather than expanded ones will nearly always be used in this book and by practicing chemists. Unless otherwise indicated, "structural formula" in this book will mean "condensed structural formula."

To name compounds we use stems that signify the number of carbon atoms present in the group of atoms being named. The stem signifying one carbon atom is <u>meth</u>, that for two carbon atoms is <u>eth</u>, that for three carbon atoms is <u>prop</u> (rhymes with <u>hope</u>), and that for four carbons is <u>but</u> (rhymes with <u>cute</u>). The stem is combined with an ending characteristic of the class of compounds. The characteristic ending for alkanes is <u>ane</u>.

5. The name for CH_4 is methane, formed by combining the stem _____, signifying one carbon

atom, and the ending _____, indicating class of compound.

6. In similar fashion, $CH_3 — CH_3$ is named _____, and $CH_3 — CH_2 — CH_3$ is named

_____.

7. A condensed structural formula for a compound named <u>butane</u> is _____.

Items 8 through 20 are concerned with the formula

$$CH_3 — CH_2 — \underset{\underset{CH_3}{|}}{CH} — \underset{\underset{\underset{\underset{CH_3}{|}}{CH_2 — CH_3}}{|}}{CH} — CH_2 — CH_3$$

8. Since this formula contains only carbons and hydrogens and has no multiple bonds between

pairs of atoms, the class of compound it represents is _____.

9. Complex alkanes can be named using the longest continuous chain of carbon atoms as the basis of the name. (<u>Note</u>: The adjective is <u>continuous</u>—not horizontal or vertical or straight, but continuous).

The longest <u>continuous</u> chain, or parent chain, of carbon atoms in the formula above contains _____
carbon atoms.   (number)

10. Draw a continuous line through the carbon atoms in this longest continuous chain.

$$CH_3 — CH_2 — \underset{\underset{CH_3}{|}}{CH} — \underset{\underset{\underset{\underset{CH_3}{|}}{CH_2 — CH_3}}{|}}{CH} — CH_2 — CH_3$$

11. Stems signifying more than four carbons in the group of atoms being named are mostly Greek (a few are Latin) in origin. For example, <u>pent</u> signifies <u>5</u>; <u>hex</u> signifies <u>6</u>; <u>hept</u>, <u>7</u>; <u>oct</u>, <u>8</u>; <u>non</u>, <u>9</u>; <u>dec</u>, <u>10</u>; and so on. The stem that signifies the number of carbon atoms in the longest continuous chain above

is _____, and the alkane name for this chain is _____.

12. All the groups attached to the chain of carbon atoms through which the line was drawn in item 10 are called substituents. There are _____ substituents shown in that formula.
_(number)

13. Substituents that resemble alkanes (that is, are composed of only carbon and hydrogen with only single bonds between pairs of atoms) are traditionally named by adding yl to the stem that signifies the number of carbon atoms in the substituent. For example, a one-carbon substituent, CH_3, is named methyl: the stem, _____, always signifies one carbon, and the ending, _____, signifies a point of attachment to something else. The stem eth always signifies _____ carbon atoms; a substituent with that number of carbon atoms is named _____. The general stem for an alkane-like grouping is alk, and the general or class name for an alkane-line substituent is _____.

14. The traditional names of the three substituents in the formula in item 10 are _____, _____, and _____.

15. The name of a compound must include a specification for each substituent. Whenever two or more of the substituents in a formula are alike, a prefix such as di (for 2) or tri (for 3) is added to the substituent name to indicate the correct multiplicity. For example, two methyl substituents will be designated not by methyl methyl, but by _____.

Substituents are cited in a name in alphabetical order (ethyl before methyl). Only the name of the substituent is used in the alphabetical ordering; a multiplying prefix (di, tri, and so forth) that is not a part of the name of the substituent is not considered when alphabetizing.

16. Substituent names precede the parent chain name as modifiers to make a single-word substitutive name. A partial name for the formula, which specifies all substituents in alphabetical order as well as the parent chain, is _____.

17. A correct name, such as that given in item 16, specifies the total number of carbon atoms in the formula and may be checked easily by comparison with the formula itself. The total number of carbon atoms shown in the formula is _____; therefore, the name must specify that same number of
_(number)

carbon atoms. The parent chain name specifies _____ carbon atoms, and the substituent names specify
_(number)

_____, _____, and _____, respectively. The name thus specifies a total of _____ carbon atoms.
_(number) _(number) _(number) _(number)

18. For a substitutive (IUPAC) name, the atoms in the parent chain are numbered from one end of the chain to the other. The direction of numbering is chosen so that the substituents are assigned the lower possible numbers (locants) designating positions along the parent chain. Each substituent must have a locant. The locants to be used for the substituents in this formula are _____, _____, and _____.

19. Locants are set off from each other by commas and from the letter part of the name by hyphens; each immediately precedes the particular substituent that it modifies. The complete name for the formula, then, is _____.

20. Let us reexamine the name. The portion <u>heptane</u> refers to _____

_____.

The portions <u>ethyl</u> and <u>dimethyl</u> refer to _____ on the parent chain; there are _____
(number)

substituents on this chain. The locants designate _____

_____.

Whenever locants occur together in a name, they are separated from each other by _____;

All locants are separated from the letter portions of the name by _____. The use of hyphens to separate locants from each other or commas to separate locants from the other parts of the name is incorrect. Punctuation is part of the grammar of chemical nomenclature and, like the grammar of other
language usage, is governed by rules intended to aid communication.

21. Now generate a name to go with the formula

$$
\begin{array}{ccccccc}
& CH_3 & & & CH_2-CH_3 & & \\
& | & & & | & & \\
CH_3-&CH-&CH_2-&CH-&CH-&CH_2-&CH_3 \\
& & & | & & & \\
& & & CH_3-CH_2 & & &
\end{array}
$$

The parent chain in this formula is the _____ chain of carbon atoms. This parent contains

_____ carbons. The stem signifying this number of carbon atoms in a group is _____, and the

alkane name for a parent compound of this many carbons is _____. There are _____ substituents
(number)

on the parent chain; these substituents are named _____, _____, and _____.

22. Numbers indicate positions along the parent chain, and it is possible to number the chain from
either end. One direction alone is usually correct, for the rule to be followed states that the substituents
must have the smaller possible locants. We compare the locants term by term and make the choice of
set at the first point of difference. For example, 1, 2, 4 will be chosen over 1, 3, 3. Numbering from

the left in the formula above assigns to the substituents the positions _____, _____, and _____.
Numbering from the other end of the parent chain would assign to the substituents the numbers

_____, _____, and _____. Since we must use the smaller possible numbers, the correct position des-

ignations are _____, _____, and _____. Each substituent must have a locant in the name; the locant

for the methyl group is _____, and the locants for the ethyl groups are _____ and _____. When two

or more locants occur together in a name, they are separated from each other by _____, and

locants are separated from the other parts of a name by _____.

23. For the formula in item 21, we are now ready to write a complete substitutive name, which is

_____.

24. The parent for the formula

$$CH_3 - \underset{\underset{CH_3}{|}}{\overset{\overset{CH_3}{|}}{CH}} - CH_2 - \underset{\underset{CH_3}{|}}{\overset{\overset{CH_3}{|}}{C}} - CH_3$$

is the chain containing _____ carbons. The alkane name for a parent compound of this many carbons
                            (number)

is _____, and the complete substitutive name for the liquid compound represented by the

formula is _____.

25. The name for the liquid compound represented by the formula

$$CH_3 - CH_2 - CH_2 - CH_2 - CH_2$$
$$CH_3 - CH_2 - CH_2 - CH_2 - \underset{\underset{CH_2 - CH_3}{|}}{\overset{|}{C}} - CH_2 - CH_3$$

is _____.

26. A structural formula for 3-ethyl-3, 4-dimethylhexane is

_____ .

27. A structural formula for 2, 4, 5-trimethyloctane is

_____ .

ISOMERS

Compounds that have the same molecular formula (which merely indicates the number and types of
atoms present in each molecule) but different structural formulas are called isomers. For example,
the compounds represented by the formulas

$$CH_3 - CH_2 - CH_2 - CH_3 \quad \text{and} \quad CH_3 - \underset{\underset{CH_3}{|}}{\overset{\overset{CH_3}{|}}{CH}} - CH_3$$

(both have the molecular formula C_4H_{10}) are isomers. The structural formulas are different: One shows
a continuous chain of four carbon atoms, but the other shows a branched chain of four carbon atoms. The
structural formulas represent two different compounds with different physical and chemical properties.

28. Compounds that may be represented by the structural formulas

$$CH_3 - CH_2 - CH_2 - CH_2 - CH_2 - CH_3 \quad \text{and}$$

$$CH_3 - CH_2 - \overset{\overset{\displaystyle CH_3}{|}}{CH} - CH_2 - CH_3$$

have molecular formulas of _____ and _____, respectively, and are called _____.

29. There are three isomers of molecular formula C_5H_{12}. The three (condensed) structural formulas are

_____, _____, and _____.

30. In each of these three structural formulas, the valence of carbon is _____, and the valence of hydrogen is _____.

31. The substitutive names of the three isomers in item 29 are _____,

_____, and _____, respectively.

Most alkane isomers containing fewer than seven carbon atoms may also be differentiated by use of structural prefixes, but <u>only</u> if there are no other substituents present. In the chemical literature, both old and current, one encounters the prefix <u>n</u>-, meaning <u>normal</u> or unbranched: for example, <u>n</u>-pentane. The IUPAC names of the unbranched isomers do not contain any structural prefix, however; the unmodified name means the unbranched isomer and is unambiguous. The archaic <u>n</u>- prefix has never had support in official rules and should never be used.

Prefixes are used for branched-chain isomers. The structural prefix <u>iso</u> signifies a single carbon branch at one end of the parent chain. The prefix <u>iso</u> is not separated from the alkane portion of the name in any way according to IUPAC rules.

32. There are two isomers with molecular formula C_4H_{10}. <u>Butane</u> is the IUPAC name for the isomer represented by the structural formula _____, and <u>isobutane</u> is the name for the isomer represented by the structural formula

_____ .

Note that the stem in both names (butane and isobutane) indicates the total number of carbons in the compound. Isobutane may also be named as a substituted alkane; the parent chain name is _____, and the substitutive name for the compound is _____.

33. The stem (to signify all the carbon atoms) to be used in the name for

$$CH_3 - CH_2 - \underset{\underset{CH_3}{|}}{CH} - CH_3$$

is _____, signifying a total of _____ carbon atoms. The appropriate structural prefix is _____,
(number)

and the complete IUPAC name using this prefix is _____.

34. The name for the compound represented by the formula

$$CH_3 - CH_2 - CH_2 - CH_2 - CH_2 - CH_3$$

is _____, and the name for its isomer,

$$CH_3 - \underset{\underset{CH_3}{|}}{CH} - CH_2 - CH_2 - CH_3$$

is _____ or _____.
(use structural prefix)   (as substituted alkane)

35. The structural prefix iso is restricted to compounds with a single carbon branch at one end of the parent chain. No structural prefix is accepted to indicate the C_6H_{14} isomer with structural formula

$$CH_3 - CH_2 - \underset{\underset{CH_3}{|}}{CH} - CH_2 - CH_3$$

This isomer must be named as a substituted pentane and will be called _____.

36. Two isomers with molecular formula C_5H_{12} are pentane and isopentane, whose structural formulas may be written, respectively,

_____ and _____.

37. A third isomer with molecular formula C_5H_{12} is known. By restricting ourselves to a valence of 4 for carbon and a valence of 1 for hydrogen, we can write only one other structural formula for C_5H_{12}. That formula is

_____.

The name of the C_5H_{12} isomer that contains one carbon bonded only to other carbons may be formed by use of the structural prefix neo. Like iso, neo is not separated from the alkane portion of the name.

38. The permissible IUPAC name with a structural prefix for the C_5H_{12} isomer described in item 37 is

_____. This prefix is not used in the specific nonsubstitutive name for any other alkane.

39. The IUPAC name for the compound represented by the formula

$$CH_3 - \overset{\overset{\displaystyle CH_3}{|}}{\underset{\underset{\displaystyle CH_3}{|}}{C}} - CH_2 - CH_3$$

is _____.

Isomers with more complex branching than that signified by the prefixes iso and neo, and alkanes containing more than six carbon atoms, are not named by use of structural prefixes. Names for these compounds are based on a substituted parent chain.

40. The IUPAC name for the compound represented by the structural formula

$$CH_3 - CH_2 - \overset{\overset{\displaystyle CH_3}{|}}{\underset{\underset{\displaystyle CH_3}{|}}{C}} - CH_2 - CH_2 - CH_3$$

is _____.

CYCLOALKANES

Cyclic hydrocarbons are named in much the same way as are acyclic ones. The operational prefix <u>cyclo</u> precedes the alkane name that would be used for a parent chain containing the same number of carbons as are present in the cycle or ring. Thus the formula

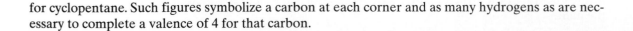

represents a liquid hydrocarbon named <u>cyclopentane</u>. Like iso and neo, the prefix <u>cyclo</u> is not separated from the alkane portion of the name in any way according to IUPAC rules. For convenience, cycloalkanes are most frequently represented by geometric figures, such as

for cyclopentane. Such figures symbolize a carbon at each corner and as many hydrogens as are necessary to complete a valence of 4 for that carbon.

41. The symbol ☐ represents a compound whose molecular formula is _____, whose structural formula using Cs and Hs is

_____ , and whose name is _____ .

42. The symbol ⬡ represents a compound named _____ .

43. Substituents on the ring are treated just as substituents on a chain for naming purposes. Thus

is named <u>methylcyclohexane</u>. The name of an isomer

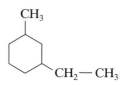

is _____ .

44. All the positions in a cycloalkane ring are equivalent, and a number is not needed to indicate the position of substitution in monosubstituted rings. If there are two or more substituents, however, locants are used to indicate positions of substitution. One of the substituents is always assigned position 1, and the smaller possible numbers (locants) are used for all others. If the same set of locants would follow the selection of different substituted positions as position number 1, the preferred selection has the first-named substituent on position 1. (Remember: Substituents are named in alphabetical order.) For example, in the name of the compound

$$CH_3$$

$$CH_2 - CH_3$$

the substituent to be named first is _____, and the number of the position to which it is attached (that is, its locant) will be _____. The locant of the other substituent will be _____, and the complete name for the compound will be _____ .

45. The name of

$$CH_3 - \diamond - CH_3$$

is _____ .

46. The use of the lowest possible set of locants always takes precedence even over the association of the first-named substituent with position 1. In the name for

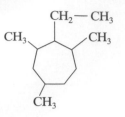

the substituent _____ is named first. If the position to which it is attached is number 1, the set of locants to be used would have to be _____. If one of the methyl groups is on position 1, however, lower sets of locants are possible. The lowest set of locants is _____, and the name of the compound is _____.

The convenience of line drawing formulas is often utilized for alkyl substituents as well as for cycloalkane rings. For example, ethylcyclopentane may be represented by the complete line drawing formula ⬠⟍ ; each corner and terminus symbolizes a carbon and as many hydrogens as are necessary to complete a valence of 4 for that carbon. If the terminus is connected to an atomic symbol, however, even in line drawings, that representation is only a bond connection and not a carbon atom; that is, ⬠⟍ is the formula for a 7-carbon compound, but ⬠⟍Cl is the formula for a 6-carbon compound with a chloro substituent.

47. A complete line drawing formula for 1,4-dimethylcyclobutane is _____,

and one for 2-ethyl-1,3,5-trimethylcycloheptane is _____.

48. The name for ⬠—Cl is _____.

2
Nomenclature of Alkyl Groups

TRADITIONAL NAMES[*]

A chemical group that can be imagined to be formed by loss of a hydrogen atom from an alkane is traditionally named by replacing <u>ane</u> of an alkane name with <u>yl</u>. The general term for such a group is <u>alkyl group</u>. The position from which the hydrogen is lost is often called the position of free valence.

1. When an alkyl group contains only one carbon, it is designated a _____ group; when it contains two carbons, it is designated an _____ group. Methyl and ethyl are definitive terms, but definitive names for higher alkyl groups must differentiate isomers.

Substitutive names are the most generally applicable and, in spite of first appearances, the simplest ones. They are used exclusively by *Chemical Abstracts* for indexing. In substitutive names of alkyl groups, the position from which the hydrogen of an alkane can be imagined to have been removed (the position of free valence) is <u>always</u> position number 1, and the longest continuous chain of carbons beginning with position number 1 is the parent chain. For example, the alkyl group $CH_3 - CH_2 - CH - CH_2 - CH_3$ is

named 1-ethylpropyl (not 1-ethyl-1-propyl and not 3-pentyl). The name 3-pentyl has about the same acceptability as the expression "I ain't."

2. The parent chain that is the basis of the substitutive name of the alkyl group

$$CH_3 - CH - CH_2 - CH_2 - \overset{|}{\underset{|}{C}} - CH_2 - CH_3$$
$$\underset{CH_3}{} \qquad \underset{CH_2 - CH_3}{}$$

contains _____ carbons and _____ substituents. The set of locants that must be used for these
(number)   (number)

substituents is _____. The complete name of this alkyl group is _____.

---

[*] An alternative IUPAC method of naming complex (substituted) alkyl groups will be included near the end of Chapter 6.

3. The compound represented by the formula

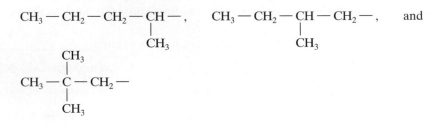

may be named as a substituted cycloalkane. The alkyl substituent is named _____

and the compound is named _____.

   Parentheses around the name of a complex substituent may be necessary for complete clarity in a name: for example (1-ethylbutyl)cycloheptane. The parentheses make clear that ethyl and butyl are parts of the same substituent rather than separate substituents.

4. The isomeric alkyl groups

$$CH_3 - CH_2 - CH_2 - CH - , \qquad CH_3 - CH_2 - CH - CH_2 - , \qquad \text{and}$$
$$\qquad\qquad\qquad\quad |\qquad\qquad\qquad\qquad\qquad\quad |$$
$$\qquad\qquad\qquad\quad CH_3\qquad\qquad\qquad\qquad\qquad CH_3$$

$$CH_3$$
$$\;\;|$$
$$CH_3 - C - CH_2 -$$
$$\;\;|$$
$$CH_3$$

have the substitutive names _____ , _____ , and _____ ,
respectively.

5. The formula

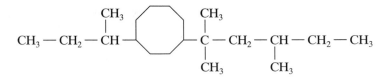

shows two substituents attached to a cyclooctane ring. The larger substituent is named

_____,and the smaller one is named _____ . The names of the two substituents appear in the complete name of the compound in alphabetical order; for complex substituents, such as those in the formula above, the initial letter of the <u>full</u> name of the substituent, even if it is part of a multiplying prefix, is the one used for alphabetizing (that is, <u>d</u> is the key letter of a substituent named 1,1-dimethyl-propyl). The first-named substituent in this compound is

_____. For a symmetrical parent ring (chain) such as cyclooctane, the first-named substituent is assigned the smaller position number (locant). The two substituents

are on positions numbered _____ and _____.The complete name for this substituted cyclooctane is

_____.

Some isomeric alkyl groups containing fewer than seven carbons may be differentiated by use of structural prefixes combined with the stem designating all the carbons in the group. These non-substitutive names are usually called <u>trivial names</u> and are also IUPAC names. In this context, "trivial" does not mean "unimportant," but it does mean "not systematic." For some of these simple groups, the trivial names have been used more frequently than have the substitutive ones, but emphasis on single names for indexing and for dependable machine searching of the literature is shifting usage in favor of the substitutive names. Trivial names are likely to be used for some time, however.

Trivial names of alkyl groups are best mastered by understanding the classification of alkyl groups. An alkyl group whose number 1 position (substitutive name numbering) is a carbon bound to only one other carbon ($CH_3 — CH_2 —$, for example) is classified as a <u>primary</u> alkyl group; one whose number 1 position is a carbon bound to two other carbons is classified as a <u>secondary</u> alkyl group; and one whose number 1 position is a carbon bound to three other carbons is classified as a <u>tertiary</u> alkyl group.

6. The stem signifying four carbons in an alkyl group is _____, and the name for the group represented by the formula $CH_3 — CH_2 — CH_2 — CH_2 —$ is _____. This group is classified as a _____ alkyl group.

7. The alkyl group represented by the formula $CH_3 — CH_2 — CH — CH_2 — CH_3$ is classified as a _____ alkyl group.

8. The alkyl group represented by the formula

$$CH_3 — CH — CH_2 —$$
$$\qquad\quad |$$
$$\qquad\quad CH_3$$

is classified as a _____ alkyl group, and its isomer,

$$\qquad\quad |$$
$$CH_3 — C — CH_3$$
$$\qquad\quad |$$
$$\qquad\quad CH_3$$

is classified as a _____ alkyl group.

9. The alkyl group represented by the formula

$$\qquad\quad CH_3$$
$$\qquad\quad |$$
$$CH_3 — C — CH_2 —$$
$$\qquad\quad |$$
$$\qquad\quad CH_3$$

is classified as a _____ alkyl group.

The structural prefixes <u>sec</u>- (for <u>secondary</u>-) and <u>tert</u>- (for <u>tertiary</u>-) may be incorporated in a name to designate a specific secondary or tertiary alkyl group, respectively, if no isomeric alkyl groups of the same classification are possible. These prefixes are underlined to indicate italics and are separated from the rest of the name by hyphens. The stem used with these prefixes specifies the total number of carbons in the alkyl group. (These specific names are restricted to the unsubstituted groups and are not to be used as parent names when substitution in the groups must be specified.)

10. The IUPAC name <u>sec</u>-butyl refers to the alkyl group represented by the structure

_____ .

No other secondary butyl group can be drawn. Because isomeric secondary alkyl groups are possible when there are more than four carbons in unbranched alkyl groups, the prefix <u>sec</u>- is not definitive and is not used for any of them.

11. There is a single tertiary butyl group, which may be represented by the structure

_____ and given the trivial name _____.

12. For the alkyl group represented by the formula

$$CH_3 - \overset{\overset{\displaystyle CH_3}{|}}{\underset{|}{C}} - CH_2 - CH_3$$

IUPAC rules permit the trivial name _____, as well as the substitutive name,

_____.

If the alkyl group has a single branch at one end and the point of attachment at the other end, the structural prefix <u>iso</u> may be used. For example,

$$CH_3 - \overset{}{\underset{\underset{\displaystyle CH_3}{|}}{CH}} - CH_2 - CH_2 -$$

may be named <u>isopentyl</u>. IUPAC usage of isoalkyl as a specific (trivial) name is restricted to alkyl groups with fewer than seven carbons. Similarly, the alkyl group related to the alkane neopentane may be named <u>neopentyl</u>. Notice that the prefixes <u>sec</u>- and <u>tert</u>- are underlined (for italics) and set off by hyphens, while the prefixes iso and neo are written without separation or underlining. This difference is an oddity of nomenclature that has little justification other than general usage. For alphabetical arrangement of substituted groups, the separated prefixes (<u>sec</u>- and <u>tert</u>-) are ignored. The initial letter of <u>sec</u>-butyl is considered to be <u>b</u>, but the initial letter of isobutyl is considered to be <u>i</u>.

13.   Isobutane is the name for the alkane represented by the structural formula

_____ ,

and isobutyl is the name of the alkyl group related to it. The isobutyl group may be represented by the structural formula

_____ and will be classified as a _____ alkyl group.
(primary, etc.)

14.   Isopentane may be represented by the structural formula

_____ ,

and isopentyl by the structural formula

_____ .

The isopentyl group is classified as a _____ alkyl group.

15. $CH_3 - CH - CH_2 - CH_2 - CH_2 -$ is named _____ .
              |
             CH_3

16.   Neopentane may be represented by the structural formula

_____ .

and neopentyl by the structural formula

_____ .

17.   The butyl groups for which the prefixes <u>sec</u>- and iso are appropriate are often confused by students. The butyl group for which the prefix <u>sec</u>- is correct contains an unbranched chain of carbon atoms; the butyl group for which the prefix iso is correct contains a _____ chain of carbon atoms.

18.   The name <u>sec</u>-butyl designates the structure

_____ .

19.   The name isobutyl designates the structure

_____ .

20.   Isobutyl is classified as a _____ alkyl group, and <u>sec</u>-butyl is classified as a _____ alkyl group.

21.   The alkyl groups named <u>sec</u>-butyl, isobutyl, and <u>tert</u>-butyl are also identified by the substitutive names _____, _____, and _____, respectively.

22.   In a departure from strictly systematic nomenclature, the group

$$CH_3 — CH — CH_3$$
$$|$$

is named isopropyl instead of <u>sec</u>-propyl. This name illustrates the concession made to familiar, well-established names when systematization was undertaken. The group

$$CH_3 — CH — CH_3$$
$$|$$

will actually be classified as a _____ alkyl group, but its IUPAC trivial name is

_____ .

23.   The compound represented by the formula

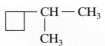

may be named as a substituted cyclobutane. The trivial name of the substituent is _____, and the name of the compound is _____ .

24.   The longest continuous chain of carbons in the formula

$$CH_3 — CH — CH_2 — CH_2 — CH — CH_2 — CH_2 — CH_2 — CH_2 — CH_3$$
$$\qquad\quad | \qquad\qquad\qquad\quad |$$
$$\qquad\quad CH_3 \qquad\quad CH_3 — CH — CH_2 — CH_3$$

contains _____ carbons. Two substituents, trivially named _____ and _____, are
(number)

attached to the parent chain at positions numbered _____ and _____, respectively. The _____
(trivial name)

substituent will be named first. The full name for the alkane is _____.

25. The alkane represented by the formula

$$CH_3 — CH_2 — CH_2 — CH_2 — CH_2 — CH — CH_2 — CH — CH_3$$
$$CH_3 — CH_2 — CH_2 — CH_2 \qquad CH_3$$

contains a parent chain of _____ carbons and a substituent whose trivial name is_____.
(number)

The substituent is located on position number _____, the name of the parent chain is _____,

and the complete name for the alkane is_____.

The IUPAC name of an unbranched primary alkyl group (for example,
$CH_3 — CH_2 — CH_2 — CH_2 — CH_2 —$, pentyl) does <u>not</u> use a structural prefix. (The structural prefix
<u>n-</u>, for <u>normal</u>, is used by some chemists, but such usage is contrary to IUPAC rules and is roughly equiv-
alent to the expression "I seen.")

26. The parent for the alkane represented by the formula

$$CH_3 — CH — CH_3$$
$$CH_3 — CH_2 \quad CH_3 — CH \qquad CH_2 — CH_3$$
$$CH_3 — CH_2 — CH — CH_2 — CH — CH_2 — CH — CH_2 — CH_3$$

is _____. The three substituents are named _____, _____, and
_____. The initial letter of the like substituents at positions 3 and 7 is _____, and that of the
one at position 5 is _____, so the one at position _____ will be named first (alphabetical order).

("Di" is part of the name of the five-carbon substituent at position 5, but "di" is <u>not</u> part of the name
of a two-carbon substituent.) The name of the branched substituent will be enclosed in parentheses
for clarity in the full name of the alkane, which is _____.

When the longest continuous chain of carbons in a formula can be identified with more than one
chain, choose as the parent the chain that has the most substituents. (This choice makes the substituents
simplest.) If alternative chains have the same number of substituents, choose the one with the simpler
(less branched) substituents or, if the substituents are alike, the one with the lower set of locants.

27. The longest continuous chain of carbons in the formula

$$CH_3 — CH_2 — CH_2 — CH — CH — CH_2 — CH — CH_2 — CH_2 — CH_3$$
$$CH_3 — CH — CH_2 \quad CH_2 \quad CH_3 — CH — CH_2 — CH_3$$
$$CH_3 \qquad CH_2 — CH_2 — CH_3$$

is _____ carbons long. There are actually six different chains with this number of carbons. The one
(number)

with the most substituents, _____, is chosen as the parent. The locants for the substituents on the
(number)

parent are _____, and the substituents are named _____,

_____, and _____. The name of the alkane represented by

the formula is _____.

   Alkyl groups may be attached to atoms other than carbon; for example, a hydrogen in an alkane
may be replaced by chlorine. These compounds may be named in two ways: <u>substitutive names</u> and <u>func-
tional class names</u>. The basis (parent compound) of a <u>substitutive name</u> can stand alone as the name of
an individual compound, and modifiers (such as substituent names) must be written as part of the same
word to avoid any ambiguity. For substitutive names, chlorine is regarded as a substituent (named <u>chloro</u>,
with locant if necessary), and the one-word substitutive name for CH_3 — Cl is <u>chloromethane</u>. (<u>Methane</u>
as a separate word could be mistaken for the name of another compound not intended.) <u>Functional
class names</u> are often two-word names, because the final portion of the name (functional class designation) cannot stand alone as the name of an individual compound. For example, CH_3 — Cl is a member
of the class alkyl chloride and is named with the functional class name, methyl chloride. (<u>Chloride</u> as a
separate word will not be mistaken for the name of a specific compound.)

28.   The trivial name of the alkyl group in the formula

$$CH_3 - CH - CH_3$$
$$|$$
$$Cl$$

is _____, the functional class name for the compound is _____,

and the substitutive name for the compound is _____.

29.   Butyl chloride may be represented by the structural formula

_____,

<u>sec</u>-butyl chloride by the formula

_____,

and <u>tert</u>-butyl chloride by the formula

_____.

30.   Neopentyl chloride may be represented by the structural formula

_____

and named with the substitutive name _____.

Neopentyl chloride is classified as a _____ alkyl chloride.
                                                      (primary, secondary, etc.)

31.   The substitutive name for

$$CH_3 - CH - CH_2 - CH_2 - Cl$$
$$\quad\quad\quad | $$
$$\quad\quad CH_3$$

is _____; the functional class name for this same compound is

_____. This alkyl chloride is classified as a _____

alkyl chloride.

32.   You should practice writing structural formulas from names of compounds. Consider the name

2,2,4-trimethylpentane. The parent chain name is _____, which signifies _____ carbons in
                                                                                    (number)

a continuous chain. Write a chain of Cs to represent this portion of the name:

_____.

There are _____ methyl substituents on the parent chain located at positions numbered _____,
              (number)

_____, and _____. Write the parent chain again, and number each of the positions:

_____ .

Now write the parent carbon chain, and place methyl groups in the proper positions.

_____ .

Complete the (condensed structural) formula by writing in Hs to indicate the normal valence of 4 for
each carbon.

_____ .

33.   In the same fashion, step by step, write a formula for 3-chloro-3-isopropyl-2,4-dimethylheptane.

_____ .

34.   There are three isomeric secondary alkyl chlorides with molecular formula $C_5H_{11}Cl$, which may be represented by the structural formulas

_____, _____, and _____.

35.   There are three isomeric tertiary alkyl chlorides with molecular formula $C_6H_{13}Cl$, which may be represented by the structural formulas

_____, _____, and _____.

Because of the existence of isomers, as illustrated in items 34 and 35, sec-pentyl chloride and tert-hexyl chloride are not definitive names for any compounds and are therefore unacceptable. Each of the six alkyl chlorides in items 34 and 35 may, however, be named with a substitutive name.

36.   A substitutive name for $CH_3 - CH - CH - CH_3$ is _____.
                                                                    $\quad\quad\quad\quad\quad\;\;\; |\quad\quad\; |$
                                                                    $\quad\quad\quad\quad\quad\;\;\; Cl\quad CH_3$

37.   2-Chloro-2,3-dimethylbutane may be represented by the structural formula

_____ .

3
Nomenclature of Alkenes

TRADITIONAL SUBSTITUTIVE NAMES

<u>Alkenes</u> contain a double bond between a pair of carbon atoms, and the systematic ending to be used in names of alkenes is <u>ene</u>. <u>Ethene</u> is a compound containing two carbons (stem <u>eth</u>) and a carbon-carbon double bond (ending <u>ene</u>). The rules for naming alkanes apply to alkenes with two additional restrictions: The chain chosen as the basis for the name is the longest continuous chain containing the carbon-carbon double bond, and the parent chain is numbered to assign that alkene linkage the smaller possible locant. The parent chain may or may not be the longest continuous chain of carbon atoms in the compound. The position of the alkene linkage is designated by the lower number assigned to the carbons joined by the double bond; this locant immediately precedes the stem of the name in traditional IUPAC names. For example, $CH_2 = CH - CH_2 - CH_3$ is named 1-butene.

1. The parent chain in the compound

$$CH_3 - CH_2 - CH - C = CH - CH_3$$
$$\underset{CH_3}{|} \quad \underset{CH_2 - CH_2 - CH_2 - CH_3}{|}$$

contains _____ carbon atoms.

2. The alkene name for this parent chain is _____.

3. The substituent on the parent chain is named _____.

4. When the parent is properly numbered, the alkene linkage is assigned the position number _____, and the substituent is assigned the position number _____.

5. The complete designation of the parent chain, including the locant for the alkene linkage, is

_____.

6. The complete name for the formula is _____.

7. The name for the parent chain in the formula

$$CH_3 - \underset{\underset{CH_3}{|}}{\overset{\overset{CH_3}{|}}{C}} - CH_2 - \underset{\overset{CH_3}{|}}{CH} - CH = CH_2$$

is _____.

8. The alkene linkage is assigned locant _____ and the substituents are assigned locants _____, _____, and _____.

9. Locants 2, 2, and 4 for the substituents are incorrect because _____

_____

_____

_____.

10. A complete name for the formula is _____.

11. On the basis of differences only in carbon skeleton and in location of the $C = C$, the formulas for the isomeric alkenes containing four carbons are

_____, _____, and _____.

12. These alkenes are named, respectively, _____, _____, and _____.

Cyclic alkenes are called <u>cycloalkenes</u>, and the alkene linkage is always assigned the locant 1. The locant is unnecessary in the name and is usually included only when substituents are present.

13. A name for

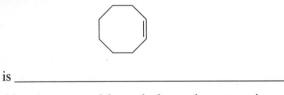

is _____.

14. A structural formula for cyclopentene is

_____.

15. A structural formula for 4-<u>tert</u>-butyl-1-cyclohexene is

_____.

16. The compound named 1-<u>sec</u>-butyl-1-cyclopentene may be represented by the formula

_____.

17. Halogens and other substituents are treated exactly as they are in names for substituted alkanes. The substituent in the formula $CH_2 = CH - CH_2 - Cl$ is located on position _____ of propene, and the substitutive name for the compound is _____.

18. The formula

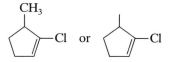

shows substituents on positions numbered _____ and _____, regardless of the direction of numbering. The choice is made to give the smaller locant (position number) to the substituent named first in the name of the compound. The compound is named _____.

   When alternative directions of numbering produce <u>different</u> sets of numbers to designate locations of substituents, the alternative sets are compared term by term. The set with the lower number at the point of first difference in the two sets is the correct one.

19. Numbering of the ring in the formula

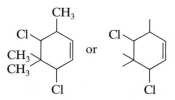

must assign numbers 1 and 2 to the carbons in the alkene linkage. If the carbon bound to chlorine is assigned position number 1, the two substituents will be located on position numbered _____ and _____. If the other alkene carbon is assigned position number 1, the substituents will be on positions numbered _____ and _____. The two sets differ in the first locant for a substituent, and the set with the smaller first locant designates positions _____ and _____. The correct substitutive name for the substituted cycloalkene is _____.

20. Substituent positions for the formula

may be designated by alternative sets of locants, which are _____ and _____. The two sets differ first in the third locant, and the set with the smaller third locant is _____. The correct name for the compound represented by the formula is _____.

NEW IUPAC NOMENCLATURE

The 1993 publication, *A Guide to IUPAC Nomenclature of Organic Compounds*, specifies, among other things, an alternative position of locants in a name: immediately before the part of the name to which it applies. For alkenes, the locant applies to the ending -ene, and the new IUPAC name for $CH_3 — CH_2 — CH = CH_2$ is but-1-ene. This style name is an alternative to the traditional name, 1-butene. Both are correct, acceptable IUPAC names for the compound. The convenience of the new alternative should become more apparent in subsequent chapters. As it becomes more familiar, this style probably will become the preferred IUPAC style among users.

21. Hex-3-ene may be represented by the structural formula _____.

22. The new IUPAC name for $CH_3 — CH_2 — CH = CH — CH_3$ is _____

and that for $CH_3 — CH — CH = CH — CH_3$ is _____.
$$\qquad\qquad\quad |$$
$$\qquad\qquad CH_3$$

23. The new IUPAC name for $CH_2 = CH — CH_2 — CH_2 — Cl$ is _____

and that for $CH_3 — CH = C — CH_2 — CH_2 — Cl$ is _____.
$$\qquad\qquad\qquad\quad |$$
$$\qquad\qquad CH_3 — CH_2$$

24. A complete line drawing formula for 1-chloro-3,3-diethyl-4,4-dimethylcyclohex-1-ene is

_____.

GROUPS RELATED TO ALKENES

An alkyl group can be considered to be formed from an alkane by loss of a hydrogen. In the same way, alkenyl groups are related to alkenes. In alkenyl groups, as in alkyl groups, the position from which the hydrogen can be considered to have been lost (the position of free valence) is traditionally always position 1, and that number does not need to be included in the name. The final e in the name of an alkene with the same parent chain as the alkenyl is replaced by the ending yl; a locant for the alkene linkage immediately precedes the stem in traditional IUPAC names. For example, $CH_2 = CH — CH_2 —$ is named 2-propenyl; the en is at position 2, and the yl is at position 1.

25. The name for the group $CH_3 — CH = CH —$ is _____.

26. Ethenyl is the name for the group _____.

Alternative, IUPAC-approved trivial names for ethenyl and 2-propenyl are vinyl and allyl, respectively. Vinyl chloride and allyl chloride are important industrial chemicals, and the trivial names are widely used.

27. The formula for vinyl chloride is _____.

28. The formula for allyl chloride is _____.

29. The allyl group will be classified as a _____ group.
_(primary, etc.)

30. The compound represented by the formula

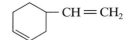

may be named as a substituted cycloalkene. The substituent, _____, is at position _____, and
_(trivial name)

the name of the compound is _____.

31. The functional class, trivial name for $CH_2 = CH — CH_2 — Br$ is _____.

Vinyl and allyl, as well as the alternative, systematic names ethenyl and 2-propenyl, respectively, may be used as the parent names for more complex alkenyl groups. The systematic names are preferred.

32. Alternative names for the alkenyl group

$$CH_2 = C — CH_2 —$$
$$\quad\quad |$$
$$\quad\quad CH_3$$

are _____ and _____.

33. The isomeric group

$$\quad\quad\quad\quad CH_3$$
$$\quad\quad\quad\quad |$$
$$CH_3 — CH = C —$$

is named _____.

34. The group 1-methyl-3-cyclopentenyl may be represented by the formula

_____ and is classified as a _____ group.
_(primary, etc.)

The new style of IUPAC nomenclature places the locant(s) adjacent to the part(s) of the name modified by the locant(s). This style of name for $CH_2 = CH — CH_2 — CH_3$ is but-1-ene and that for $CH_2 = CH — CH_2 — CH_2 —$ is but-3-en-l-yl. (The latter name illustrates the rule that preference for low locant goes to the structural fragment of higher priority; in this case yl over ene. It also illustrates the current preference for including the locant 1 when other locants appear in the name.

35. The new style name for $CH_3 — C = CH — CH — CH_2 —$ is _____.
$$\quad\quad\quad\quad\quad | \quad\quad\quad\quad |$$
$$\quad\quad\quad\quad\quad CH_3 \quad\quad CH_3$$

In line drawings, such as those used for cycloalkanes and cycloalkenes, the position of free valance (yl) must be symbolized in a way that avoids confusion with a methyl group. The IUPAC-recommended symbol is a wavy line at the end of and perpendicular to a straight line: is a formula for cyclohex-1-en-1-yl.

36. The new style name for the substituent is _____

and that for the compound

$$CH_2 - CH_2 - CH_2 - CH_2 - CH_2 - CH_3$$

$$CH - CH_2 - CH_3$$

is _____.

The 4 outside the parentheses refers to a position on _____, and the 4 inside the

parentheses refers to a position on _____.

CIS, TRANS ISOMERS

37. On the basis of differences only in carbon skeleton and in location of the $C=C$, the formulas for the isomeric alkenes containing four carbon atoms are

_____, _____, and _____.

Because of the restriction of rotation about a carbon-carbon double bond, some alkenes can exist as spatial isomers. The distinction between these isomers depends on the spatial arrangement of the groups attached to the $C=C$ rather than on the position of substituents or the alkene linkage along the parent chain. Such isomerism is one kind of stereoisomerism; the designation <u>cis,trans isomerism</u> is often used. For example,

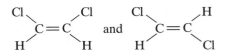

are cis,trans isomers. The two atoms or groups attached to <u>each</u> carbon of the alkene linkage must be different for cis,trans isomerism to be possible. That is, if one of the alkene carbons is attached to two identical atoms or groups, cis,trans isomerism will not be possible for that compound.

38. Of the alkenes represented by the formulas in item 37, only _____ can exhibit

(number)

cis,trans isomerism, namely _____.

39. Formulas that reveal the cis,trans isomerism of but-2-ene are

_____ and _____.

The adjectives <u>cis</u> and <u>trans</u> are used to differentiate cis,trans isomers. Cis designates the isomer with reference groups or atoms on the same side (side, not end) of the alkene linkage; trans designates the isomer with reference groups or atoms on opposite sides of the alkene linkage. Cis and trans designations for alkenes refer to the extension of the parent chain from the alkene linkage.

40. The cis isomer of but-2-ene may be represented by the formula

_____.

41. The trans isomer of but-2-ene may be represented by the formula

_____.

The prefixes <u>cis</u>- and <u>trans</u>- are included in names that specify the spatial arrangement (configuration) of alkenes. These italicized prefixes are separated from the rest of the name by hyphens, and they immediately precede the locant of the alkene linkage in traditional names or the stem of the alkene name in the new style names. (When cis and trans are used as adjectives and not as prefixes in names, they are not italicized in general use.)

42. The structural formula for <u>trans</u>-pent-2-ene is

_____ .

43. The prefix <u>trans</u>- specifies that _____.

44. A complete name that specifies configuration for

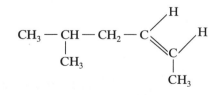

is _____.

45. A structural formula for 3,4-dichloro-9-methyl-<u>trans</u>-3-decene is

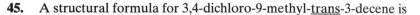

_____.

Cycloalkenes with ring sizes smaller than eight members have been isolated only as <u>cis</u>-cy-cloalkenes. Cis,trans isomers of cycloalkenes with eight-membered rings and larger have been isolat-ed. For convenience, cycloalkenes are usually represented by geometric figures, as are cycloalkanes. A cis alkene linkage is represented by _/ (extensions of parent chain from the same side of $C = C$), and a trans alkene linkage by (extensions of parent chain from opposite sides).

46. Cyclooctane may be represented by the formula . In a line drawing (geometric fig-ure) formula for each of the two cyclooctenes, the fragment of the formula that represents the cis alkene linkage will look like _____, and the fragment that represents the trans alkene link-age will look like _____. <u>cis</u>-Cyclooctene may be represented by the formula

_____,

and <u>trans</u>-cyclooctene by the formula

_____.

47. The geometry indicated for the alkene linkage by the formula

is _____. The ring contains _____ carbons, and a complete name for the compound illustrated is
      (cis or trans)   (number)

_____.

The configuration (spatial arrangement) of some isomers, such as

$$CH_3\diagdown \quad \diagup Cl$$
$$C=C \quad \text{and}$$
$$CH_3 - CH_2 \diagup \quad \diagdown Br$$

$$CH_3\diagdown \quad \diagup Br$$
$$C=C$$
$$CH_3 - CH_2 \diagup \quad \diagdown Cl$$

is not readily specified by the labels cis and trans. A more general specification of configuration of alkenes has been devised and is widely used, even in place of cis and trans with simple alkenes. This specification depends on assigning higher or lower priority to each of the two atoms or groups bound to each carbon in the alkene linkage. If the two higher-priority atoms or groups are on the same side of the C=C, the label (prefix) is <u>Z</u> (for German <u>zusammen</u> = together); if they are on opposite sides, the label (prefix) is <u>E</u> (for German <u>entgegen</u> = opposite).

Priorities of the atoms or groups in each pair are assigned by application of a comprehensive set of rules. Three rules selected from that set are likely to cover most alkenes in a beginning course.

(1) The higher the atomic number of the atom directly attached to the alkene carbon, the higher the priority of that atom (or group). For examples, CH_3 — is higher priority than H — (atomic number 6 vs. 1), but Cl — is higher than CH_3 — (17 vs. 6).

(2) For isotopes of the same element, the higher priority goes to the one of higher mass number [^2H (or D) higher than ^1H].

(3) If the two atoms attached directly to the alkene carbon are the same, the one attached in turn to atoms of higher atomic number has the higher priority. For example, CH_3 — CH_2 — is higher priority than CH_3 — (C attached C,H,H vs. C attached to H,H,H), but $(CH_3)_2CH$ — is higher priority than CH_3 — CH_2 — (C attached to C,C,H vs. C attached to C,H,H).

In names, the prefix E or Z appears first, is underlined (italicized), enclosed in parentheses, and separated by a hyphen from the rest of the name: for example, (E)-2-pentene or (E)-pent-2-ene.

48. In the formula

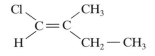

the two atoms attached to the left carbon in the alkene linkage (C $=$ C) have the priorities: higher

_____, lower _____. This priority order is assigned on the basis of _____

_____.

The two atoms attached to the right carbon in the C $=$ C are the same, but priorities of the groups can

be assigned on the basis of _____

_____.

For those groups, the priorities are: higher _____, lower _____. The two higher-priority

atoms/groups appear on the _____ side(s) of the C $=$ C, and the configuration

of this alkene is _____. The parent alkene of this compound is named _____,
                   (E or Z)

and the full name of the compound illustrated, including specification of configuration, is

_____.

49. The configuration of

$$CH_3 - CH_2 \diagdown \qquad \diagup CH_2 - CH_3$$
$$C = C$$
$$CH_3 - CH_2 - CH \diagup \qquad \diagdown CH_3$$
$$\vert$$
$$CH_3$$

may be specified as E or Z by assigning priorities to the two pairs of alkyl groups attached to the alkene carbons. Besides an alkene carbon, the first carbon in each alkyl group is attached in turn to three other atoms. In ethyl, they are _____, _____, and _____; in sec-butyl, _____, _____, and

_____; and in methyl, _____, _____, and _____. Ethyl has _____ priority than does sec-
                                                                   (higher or lower)
butyl and _____ priority than does methyl. The configuration of this alkene is therefore
            (higher or lower)
_____, and the full name for the alkene is _____.

If the alkene were designated cis or trans, the correct designation would be _____.

50. The configuration of the alkene represented by the formula

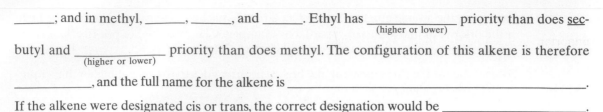

is _____ or _____. (Note that E, Z and cis,trans here and in item 49 come from the use of
   (E or Z)   (cis or trans)

different reference groups, that E and trans are not always associated with the same structure, and that
the prefixes would appear at different points in names of the alkenes.)

51. (Z)-2—Chloro-1-cyclopropylbut-1-ene may be represented by the formula

_____.

52. The compound

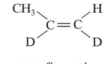

has the _____ configuration.
        (E or Z)

DIENES

Hydrocarbons containing two separate carbon-carbon double bonds are named as alkadienes. The
appropriate stem signifying the number of carbon atoms in the parent chain is combined with the
ending adiene. To learn why the "a" is included in the ending, try saying aloud "pentdiene" and
"pentadiene." The "a" simply makes the word easier to pronounce. The parent chain is numbered
to give the lower possible locants to the double bonds; a locant for each double bond precedes the
stem of the name in traditional style names or the diene ending in new style names.

53. The alkadiene with the traditional name 1,4-pentadiene may also be named _____

_____ (new IUPAC style) and may be represented by the structural formula

_____. Its isomer, 1,3-pentadiene (also correctly named _____)

may be represented by the structural formula _____.

54.   The hydrocarbon represented by the formula $CH_2 = CH - CH = CH_2$ is probably the most important diene from the standpoint of industrial use; it is named _____ or
(traditional)

_____ .
(new)

55.   Isoprene is the common (trivial) name accepted by IUPAC for 2-methylbuta-1,3-diene, which may be represented by the formula

_____ .

Isoprene is regarded as the building block for many complex compounds occurring in nature.

56.   A structural formula for 6-chloro-8-methyl-3-propyl-2-<u>cis</u>,6-<u>trans</u>-nona-2,6-diene is

_____. This diene is classified as a(n) _____ diene.

57.   Propadiene, _____, is also known as allene. Allene is accepted as an IUPAC name for
(structural formula)

this unsubstituted diene, but the more systematic name, propadiene, is preferred.

58.   The name of the compound $CH_3 - CH = C = CH_2$ is _____ .

59.   1,3-Cyclohexadiene (cyclohexa-1,3-diene) may be represented by the formula

_____ .

1,2-Dienes are classified as cumulated dienes, 1,3-dienes as conjugated dienes, and dienes with greater separation of the two double bonds as isolated dienes.

60.   Isoprene (2-methybuta-1,3-diene) is classified as a(n) _____ diene.

61.   The hydrocarbon represented by the formula

is classified as a(n) _____ diene.

62.   Propadiene, represented by the structural formula _____, is classified as a(n) _____ diene.

63.   2-Chloro-5-ethyldeca-3,5-diene, represented by the structural formula

_____ ,

is classified as a(n) _____ diene. A formula that illustrates the ($3\underline{Z},5\underline{Z}$) configuration for

this diene is _____.

(<u>Note</u>: For priority assessment, each bond from C is considered separately; the group fragment $C = C$ is considered as C attached to C,C.)

Hydrocarbons containing more than two separate carbon-carbon double bonds are named in similar fashion, <u>di</u> being replaced by the appropriate multiplying prefix.

64.   Cycloocta-1,3,5,7-tetraene has _____ $C = C$, each of which is cis; the compound can be

(number)

represented by the line-drawing formula

_____.

BIVALENT GROUPS

In the same way that one can imagine the formation of a univalent group (alkyl) by removal of a hydrogen from an alkane, one can imagine the formation of a bivalent group by removal of two hydrogens from an alkane. As with the names of alkyl and alkenyl groups, the names of bivalent groups may be used in functional class names of compounds. Except for rather simple structures, however, such names are seldom encountered. The simplest bivalent group, $— CH_2 —$, is named <u>methylene</u>, and CH_2Cl_2 is often named <u>methylene dichloride</u>.

65.   The bivalent group, $— CH_2 — CH_2 —$, may be called <u>dimethylene</u> or <u>ethylene</u>. Ethylene dibromide is an acceptable IUPAC name for $Br — CH_2 — CH_2 — Br$, but that compound is indexed

in *Chemical Abstracts* only by its substitutive name,_____.

Functional class names with these bivalent names offer no advantage over the systematic, substitutive names, and their use seems to be decreasing.

4
Alkynes

Alkynes contain a triple bond between a pair of carbon atoms somewhere in the molecule and are named in much the same way as are alkenes. The systematic ending yne, signifying the triple bond between a pair of carbon atoms in the parent chain, is combined with the appropriate stem, and the position of the $C \equiv C$ is indicated by the smaller possible locant, which traditionally precedes the stem of the name. In the new IUPAC style, this locant immediately precedes the ending yne.

1. The name for $HC \equiv CH$ is ethyne, and the name for $CH_3 - C \equiv CH$ is _____. Ethyne is also known as acetylene, an alternative parent name accepted by IUPAC. There is little other than familiarity to recommend usage of this older name, however, and *Chemical Abstracts* does not use it for indexing. Ethyne is shorter and systematic.

2. In the formula

$$CH_3 - CH - C \equiv C - CH - CH_2 - CH_3$$
$$\quad\quad\quad | \quad\quad\quad\quad\quad\quad\quad |$$
$$\quad\quad\quad CH_3 \quad\quad\quad\quad CH_2 - CH_3$$

the longest continuous chain containing the alkyne linkage, $C \equiv C$, contains _____ carbons; the

(number)

stem signifying this number of carbons is _____. The ending signifying the alkyne linkage is _____,

and the name for the chain containing this group is _____. When the chain is properly

numbered, the alkyne linkage will be assigned position number _____, and the traditional parent

name, including locant, will be _____. The two substituents, _____ and _____,

are on positions _____ and _____, respectively, and the traditional full name for the compound

represented by the formula is _____.

3. The new style name for $CH_3 - C \equiv C - CH_3$ is _____.

4. The structural formula for 3-hexyne is _____.

5. The structural formula for 4-sec-butyl-1-chloro-2-octyne will contain a parent chain of _____

(number)

carbons, signified by the stem _____. The ending, yne, indicates the group _____. The parent chain can

then be written _____. The substituent on position 1 is represented by the symbol _____, the one on position 4 by

_____.

The complete structural formula for 4-<u>sec</u>-butyl-1-chlorooct-2-yne may be drawn as

_____. The systematic, substitutive name for the substituent on

position 4 is _____, and the name for the compound illustrated, using this substituent

name, is _____.

MULTIPLE MULTIPLE BONDS

Compounds containing more than one C≡C may be named as <u>alkadiynes</u>, <u>alkatriynes</u>, and so forth. The names parallel those you have studied for alkadienes.

6. 3-Isobutylhexa-1,4-diyne may be represented by the formula

_____. When the systematic, substitutive name is used for the substituent, this compound will be named

_____.

7. The parent compound on which the name of

$$CH_2 - C \equiv C - CH - CH_3$$
$$CH_3 - CH$$
$$CH - CH_3$$
$$CH_2 - C \equiv C - CH_2$$

will be based is _____. Correct numbering of the

parent ring will assign to the substituted positions the numbers _____, _____, and _____.

The complete name of the compound illustrated is _____.

IUPAC substitutive names for compounds containing both a $C=C$ and a $C\equiv C$ are based on the general parent name "alkenyne." The parent chain is numbered so as to use the lower locants for the double and triple bonds. For example, $CH_3-CH=CH-C\equiv CH$ is named pent-3-en-1-yne, not pent-2-en-4-yne. [In traditional IUPAC names, the locant for $C=C$ (ene) precedes the stem and that for $C\equiv C$ (yne) precedes the yne ending. This arrangement is used in traditional names for compounds in other functional group classes (such as alcohols, Chapter 6) that also contain carbon-carbon multiple bonds.] In names that include different endings (suffixes) the convenience of the new style, with locant directly associated with the part of the name it modifies, is apparent. If two directions of numbering the parent chain of an alkenyne use the same locants for the two multiple-bond groups, the lower number for the double bond is used, even though yne is the final ending of the name. For example, $CH_3-CH=CH-CH_2-C\equiv C-CH_3$ is named hept-2-en-5-yne (traditional style, 2-hepten-5-yne), not hept-5-en-2-yne (traditional style, 5-hepten-2-yne).

8. The compound $CH_2=CH-C\equiv CH$ is an important industrial chemical that is commonly called <u>vinylacetylene</u>. IUPAC does not approve the use of "acetylene" as the parent on which a substitutive name is built, and the IUPAC-approved name for this compound is _____.

9. The traditional name of the compound represented by the formula

$$\begin{array}{ccccccc} & & Cl & & CH_3-CH-CH_3 & & \\ & & | & & | & & \\ CH_3-C\equiv C-CH-CH_2-C=CH-CH_2-CH_3 \end{array}$$

is _____,

and that of the isomer,

$$\begin{array}{ccccccc} & & Cl & & CH_3-CH-CH_3 & & \\ & & | & & | & & \\ CH_3-C\equiv C-CH-CH_2-CH-CH=CH-CH_3 \end{array}$$

is _____.

10. The new style name of

$$\begin{array}{c} CH_3 \\ | \\ CH_2=CH-CH_2-C-C\equiv CH \\ | \\ CH_2-CH_3 \end{array}$$

is _____.

ALKYNYL GROUPS

The formulas and names of alkynyl groups are related to those of alkynes in the same way as formulas and names of alkenyl groups are related to those of alkenes.

11. The group $HC\equiv C-$ is often regarded as a substituent and is named _____.

12.   1-<u>tert</u>-Butyl-4-ethynylcyclohex-1-ene may be represented by the formula

_____ .

13.   The formula for the 2-propynyl group is _____, and that for the 1-propynyl group is _____. Remember: The point of attachment of the alkyl, alkenyl, or alkynyl group is always the number 1 position of that group in traditional names.

14.   Three isomeric groups can be formed by the removal of one hydrogen from 1-penten-3-yne. The formulas for these three groups are _____, _____, and _____, and their traditional names are _____, _____, and _____, respectively.

5
Aromatic Hydrocarbons

Benzene, the parent aromatic hydrocarbon, is symbolized in various ways:

C_6H_6 ,    ,

The orientation of the hexagon does not matter.

Benzene serves as the basis of the substitutive names of many substituted benzenes. R is often used as a general symbol for an alkyl group. The one-word name for an alkyl-substituted benzene such as

⬡— R   or C_6H_5 — R

is formed by following the name of the alkyl group with benzene (for example, methylbenzene). Since all positions in benzene are equivalent, no number is needed in the name to indicate the position of a single substituent.

1. The aromatic hydrocarbon represented by the formula C_6H_5—CH_2—CH_3 or

⬡— CH_2 — CH_3

will be named _____ .

2. The ethynyl group may be represented by the formula _____, and ethynylbenzene may be represented by the formula

_____ .

3. sec-Butylbenzene may be represented by the formula

_____ .

4. The alkyl group attached to C_6H_5 — in the formula

$$C_6H_5 - CH_2 - \overset{\displaystyle CH_3}{\underset{\displaystyle CH_3}{\overset{|}{\underset{|}{C}}}} - CH_3$$

has the trivial name _____, and the substituted benzene is named

_____.

The class of compounds of which benzene and substituted benzenes are members is often called <u>arene</u>. This general term is related to benzene as alkane is related to methane.

Sometimes, when another fragment of the molecule has priority as the parent compound, the benzene portion of the molecule is regarded as a substituent rather than as the parent compound. As a substituent, C_6H_5 — or

is named <u>phenyl</u> and is treated as any other substituent. (The completely systematic name is benzenyl, but this name has never challenged phenyl in usage.) Phenyl is an example of an <u>aryl</u> substituent, as methyl is an example of an alkyl substituent. For convenience, phenyl is frequently symbolized ϕ or Ph.

5. $(C_6H_5)_3CH$ is most easily named as a substituted methane; its name then is

_____.

6. Ethynylbenzene, represented by the formula

_____ ,

may also be named as a substituted alkyne. The name on that basis will be _____.
<u>Names based on the larger parent compound are generally preferred</u>, but occasionally, when a list of related compounds is being compiled, for example, the other choice may be acceptable.

7. An IUPAC name for

$$CH_3 - CH = CH - \underset{\displaystyle \underset{|}{CH_3}}{CH} - CH_2 - \underset{\displaystyle \underset{|}{CH_2 - CH_2 - CH_3}}{CH} - \phi$$

with phenyl being treated as a substituent like methyl, will use as the basis of the name a hydrocarbon

chain named _____. The methyl substituent appears on position number _____, and

the phenyl substituent on position number _____. A complete IUPAC name for the substituted

alkene is _____.

8. A formula for 5-phenyldodecane (dodec = 12) may be drawn as

_____.

9. Allylbenzene, represented by the formula _____, may be named as a substituted

alkene: _____. Its isomer, (1-propenyl)benzene, represented by the formula

_____, may also be named as a substituted alkene: _____.

    When two or more substituents are attached to the benzene ring, isomers are possible, and position designations must be used. Two different position designations are used: numbers and letters.

    When numbers are used, one of the substituted positions will always be numbered 1, and the other positions in the ring are numbered 2 through 6. The number 1 position and the direction of numbering are chosen so that the set of lowest numbers is assigned to the substituted positions. (The set of lowest locants has the lowest number at the first point of difference.)

10. In the formula

$$CH_3 - \bigcirc - CH_3$$

the methyl substituents appear on positions numbered _____ and _____.

11. 1,2-Diethylbenzene may be represented by the formula

_____.

12. The formula

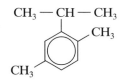

pictures substituents on positions numbered _____, _____, and _____ (the lowest numbers that can be used). The name of the compound represented by this formula is _____.

13. Mesitylene is a common name sometimes used for 1,3,5-trimethylbenzene,

_____.

    (formula)

For many disubstituted benzenes, chemists frequently use letters rather than numbers to designate positions of substitution: o- (standing for and read <u>ortho</u>-) is used for compounds in which substituents appear on positions numbered 1 and 2; <u>m</u>- (for <u>meta</u>-) signifies substituents on positions 1 and 3; and <u>p</u>- (for <u>para</u>-) signifies substituents on positions numbered 1 and 4. These letter designations are prefixes that are italicized and set off from the rest of the names by hyphens.

14.   Xylene is a common name for dimethylbenzene. There are three isomeric xylenes known:

<u>o</u>-xylene, _____ ,
                        (formula)

<u>m</u>-xylene, _____ , and
                        (formula)

<u>p</u>-xylene, _____ .
                        (formula)

15.   1,4-Diisopropylbenzene may also be called _____.

16.   The two alkyl substituents in the formula

$$\text{C}_6\text{H}_4 \begin{cases} \text{CH}_2 - \text{CH}_3 \\ \text{CH} - \text{CH} - \text{CH}_3 \\ \quad | \qquad | \\ \quad \text{CH}_3 \quad \text{CH}_3 \end{cases}$$

are _____ to each other.
        (o-, m-, or p-)

17.   <u>m</u>-Ethylvinylbenzene can be represented by the formula

_____ .

18.   1-Hexyl-2-isobutylbenzene may also be named _____.
                                                              [letter(s) as locant(s)]

Methylbenzene is usually called <u>toluene</u> (even though *Chemical Abstracts* indexes the compound as methylbenzene), and <u>toluene</u> is actually used as the basis of IUPAC names. That is, toluene is treated as a parent compound just as benzene is. When <u>toluene</u> is used in this way, the carbon to which the methyl group is attached is position number 1.

19.   3-Ethyltoluene may be represented by the formula

_____ .

20.   p-<u>tert</u>-Butyltoluene may be represented by the formula

_____ .

21.   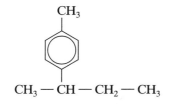 $CH_2 — CH = CH_2$ may be named as a substituted toluene, _____.

22.   Named as a substituted benzene,

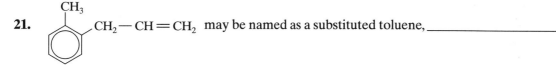

will be called _____; named as a substituted toluene,

the same compound will be called _____.

   Some other substituted benzenes are also treated as parent compounds like toluene.

23   For example, ethenylbenzene, represented by the formula

_____ ,

is most often called <u>styrene</u>. *Chemical Abstracts*, however, uses only ethenylbenzene for indexing purposes.

24.   p-Isobutylstyrene may be represented by the formula

_____ .

The two positions in the vinyl group of styrene have been designated by the Greek letters <u>alpha</u> and <u>beta</u>, as illustrated in the following formula:

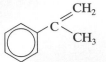

This style of position designation is no longer recommended. Names based on benzene as the parent compound are preferred and are the ones used by *Chemical Abstracts*.

25. The compound represented by the formula

has been called α-methylstyrene, but it should be named as a substituted benzene,

_____ ;

an isomer, (1-propenyl)benzene, may be represented by the formula

_____ ,

another isomer, 3-methylstyrene, may be represented by the formula

_____ .

26. The substituted styrene

may be named as a substituted alkene with the name _____ ; or it may be named as a substituted benzene (*Chemical Abstracts*). The choice of name often depends upon the context in which one wishes to use the name.

27. Phenylbenzene, represented by the formula

_____ ,

is usually called biphenyl. <u>Biphenyl</u> is an IUPAC-approved parent compound name. The positions in each ring are numbered 1 through 6, beginning at the carbon bonded to the other ring; one set of locants is primed (1' through 6') to distinguish that ring from the other. For example,

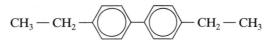

is named 4,4'-diethylbiphenyl. The unprimed locants have higher priority; that is, they will be associated with the ring having more substituents.

28. The compound

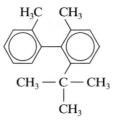

is named _____.

A polycyclic arene that is the parent compound of a number of industrial products is naphthalene, $C_{10}H_8$,

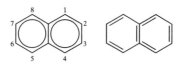

The locants for the positions are shown in the formula. The aryl groups related to naphthalene,

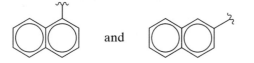

and

are traditionally named 1-naphthyl and 2-naphthyl (*Chemical Abstracts* uses the longer names 1- and 2-naphthalenyl). Note that established, invariant numbering of polycyclic compounds such as naphthalene permits and requires a locant other than 1 for some of the related groups.

29. 1-Methylnaphthalene may be represented by the formula

_____,

2-phenylnaphthalene by the formula

_____,

and 1,7-dimethylnaphthalene by the formula

_____.

30. The compound

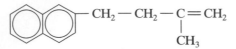

may be named as an alkene or as a naphthalene. As an alkene, it will be named _____
_____, and as a naphthalene, it will be named _____
_____, which is usually the preferred name (larger parent compound).

Hydrogens added to an unsaturated parent compound (such as an arene) may be indicated in a substituted name (based on the parent compound) by the prefix <u>hydro</u> together with the appropriate multiplying prefix and locants. (<u>Hydro</u> means "hydrogen added" rather than "hydrogen substituted.") For example,

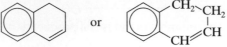

is related to naphthalene by the addition of hydrogen at positions 1 and 2 and is usually named 1,2-dihydronaphthalene. Note that the components of the name ($2H + C_{10}H_8$) add up to the correct composition of the compound ($C_{10}H_{10}$.)

31. [structure] , related to naphthalene by addition of hydrogens at positions _____ and _____,

may be named _____; [structure] , related to naphthalene by

addition of hydrogens at position _____, may be named _____;

and [structure] may be named _____.

The hydro prefix does not mean substitution, so it does not join the alphabetical ordering of substituent prefixes. Rather, it remains attached to the part of the name that it modifies and has priority over substituent prefixes (but not over suffixes) for lowest locants.

32. A structural formula for 5-methyl-1,2-dihydronaphthalene is

_____ .

33. A structural formula for 1-chloro-4-ethyl-1,4-dihydronaphthalene is

_____.

34.   A substitutive name for 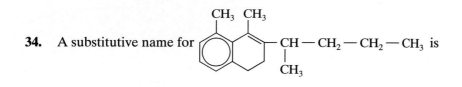 is

_____.

6
Alcohols

Alcohols are compounds in which an alkyl group is attached to an —OH group and may be represented by the general formula R—O—H (R is a convenient symbol for an unspecified hydrocarbon group). IUPAC names for alcohols are usually one-word substitutive names or two-word functional class names. Both kinds of names are used by chemists, especially for some of the low-molecular-weight alcohols for which the functional class name may be the more familiar one. *Chemical Abstracts*, with its emphasis on indexing and a single name for a compound, however, does not use the functional class names for indexing.

FUNCTIONAL CLASS NAMES

The functional class of organic compounds with the general formula R—OH is alcohol. Alcohol is a classification term, not the name of any individual compound, and it is treated in much the same fashion as the classification term chloride. The name of the alkyl group attached to —OH is specified, and the name of the compound is completed by the separate word alcohol. For example, CH_3—OH is methyl alcohol.

1. The functional class name for

$$CH_3 - \underset{\underset{OH}{|}}{CH} - CH_3$$

is _____.

2. The structural formula for sec-butyl alcohol is

_____ .

3. Functional class names of alcohols are two-word names because _____
_____.

4. The trivial name for the group

$$CH_3 - \underset{\underset{CH_3}{|}}{CH} - CH_2 - CH_2 -$$

is _____, and the functional class name for the compound represented by the formula

46

$$CH_3 - CH - CH_2 - CH_2 - OH$$
$$\vert$$
$$CH_3$$

is _____ .

5.   The IUPAC trivial name for the group $CH_2 = CH - CH_2 -$ is _____, and the functional class name for $CH_2 = CH - CH_2 - OH$ is _____ .

6.   The condensed structural formula for <u>tert</u>-butyl alcohol is

_____ .

On the same basis as is used for alkyl chlorides, alcohols can be classified as primary, secondary, and tertiary. For example, if the carbon attached to the OH group is attached to only one carbon, the alcohol is a primary alcohol.

7.   Ethyl alcohol, represented by the structural formula

_____ , is classified as a _____ alcohol.

8.   Isobutyl alcohol, represented by the structural formula

_____ , is classified as a _____ alcohol.

9.   Allyl alcohol is a _____ alcohol.

10. These three alcohols are all primary alcohols because _____

_____

_____ .

11. Isopropyl alcohol is classified as a _____ alcohol.

12. A secondary alcohol whose molecular formula is C_4H_9OH has the structural formula

_____ and the functional class name _____ .

13. There are three secondary alcohols with molecular formula $C_5H_{11}OH$; their structural formulas are

_____ , _____ , and _____ .

14. The tertiary alcohol of lowest molecular weight has the structural formula

_____ and the functional class name _____ .

15. $C_6H_5 - CH_2 - OH$ is classified as a _____ alcohol.

(primary, etc.)

SUBSTITUTIVE NAMES

Substitutive names for alcohols illustrate the pattern that is used for substitutive names of other functional group classes: The hydrocarbon name (parent compound name) for the longest continuous chain containing the functional group is modified by a systematic <u>suffix</u> that signifies the functional group. A locant to specify location of the functional group on the parent chain either immediately precedes the stem of the parent chain name (traditional) or immediately precedes the suffix (new style). The parent chain is numbered so that the functional group signified by the suffix has the lower possible locant.

    The suffix that signifies alcohol or OH group is <u>ol</u>. Whenever the suffix begins with a vowel, the final <u>e</u> in the name of the parent compound is elided (dropped). For example, the IUPAC substitutive name for

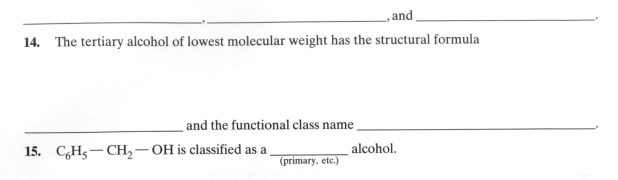

is 2-propanol or propan-2-ol. Note that the name isopropanol for this (or any) compound is incorrect, because there is no parent alkane, isopropane.

16. The traditional IUPAC substitutive name for

$$CH_3 - CH_2 - \underset{\underset{OH}{|}}{CH} - CH_2 - CH_3$$

is _____ , and the new one is _____ .

17. The condensed structural formula for 2-butanol is

_____ .

18. The systematic name for

is _____.

Substituents are named and numbered as in other IUPAC names. Remember that the OH group is assigned the lower possible locant.

19. The substitutive name for

$$CH_3 - CH - CH - CH_3$$
$$\qquad | \qquad |$$
$$\qquad OH \quad CH_3$$

is _____ or _____.
   (traditional)   (new)

A complex structural formula can usually be named rather easily in steps. Items 20 through 26 apply to the formula

$$
\begin{array}{c}
\qquad\qquad Cl \quad\; C_6H_5 \\
\qquad\qquad | \qquad\; | \\
CH_3 - CH_2 - CH - C - CH_2 - CH - CH_2 - CH_2 - CH_3 \\
\qquad\qquad\qquad | \qquad\qquad | \\
\qquad\qquad\qquad OH \qquad\; CH - CH_3 \\
\qquad\qquad\qquad\qquad\qquad | \\
\qquad\qquad\qquad\qquad\quad CH_3
\end{array}
$$

and develop the name for it in steps.

20. The longest continuous chain of carbon atoms containing the functional group that will serve as the basis of the name contains _____ carbon atoms. The IUPAC name for an alkane containing this
                    (number)
number of carbon atoms is _____.

21. The IUPAC substitutive (parent) name for an alcohol with this many carbon atoms is

_____.

22. When the parent chain is properly numbered, the OH group will be on the carbon atom numbered _____. Therefore, the parent compound name, including locant, will be _____
                                                                                              (traditional)
or _____.
         (new)

23. There are _____ substituents on the parent chain, besides the functional group; these sub-
                (number)
stituents are named _____, _____, and _____; they are located on carbons
numbered _____, _____, and _____, respectively.

24. As in all IUPAC names, the locants are separated from the letter parts of the name by _____ (and from each other, if necessary, by _____).

25. The complete traditional IUPAC name for the complex formula above then becomes

_____ .

26. This compound can be classified as a _____ alcohol.

27. Phenylmethyl alcohol, represented by the formula _____ , may also be given the substitutive name _____ . For indexing convenience, *Chemical Abstracts* uses for this compound the name benzenemethanol—an example of another IUPAC style of nomenclature, conjunctive names. Even if one does not find this style of name in general use, there is little or no difficulty in translating such a name in *Chemical Abstracts* into the intended structure.

ALCOHOLS WITH TWO OR MORE FUNCTIONAL GROUPS

Some alcohols contain two or more OH groups. Substitutive names for these compounds simply include a locant for each OH group and the appropriate ending <u>diol</u>, <u>triol</u>, <u>tetrol</u>, etc. For example, $HO-CH_2-CH_2-OH$ may be named 1,2-ethanediol. Note that the final "e" of the alkane portion of the name is retained for <u>alkanediol</u> but is omitted for alkanol. In general, the final "e" is retained with systematic suffixes beginning with consonants and dropped with those beginning with vowels.

Some diols of low molecular weight are often named by two-word functional class names that use the class designation <u>glycol</u>, together with the bivalent hydrocarbon group name. For example, $HO-CH_2-CH_2-OH$ is commonly called <u>ethylene glycol</u>.

28. Propane-1,2-diol may be represented by the formula

_____ and also identified by the functional class name, propylene glycol.

29. Its isomer, $HO-CH_2-CH_2-CH_2-OH$, has the substitutive name _____ , and the functional class name, trimethylene glycol. In general, substitutive names for diols are preferred over the glycol names.

30. The new style substitutive name for the compound

$$HO-CH_2-CH-CH_2-OH$$
$$|$$
$$OH$$

is _____ . This compound is commonly called <u>glycerol</u>.

31. The parent chain in the formula

$$\underset{CH_3}{\overset{\overset{\displaystyle OH}{|}}{CH_3-CH-CH_2-CH-CH-C-C_6H_5}}$$

is numbered so that the OH groups are assigned positions _____ and _____. The rule dictating the direction of numbering requires that the OH groups be assigned the _____ locants. The IUPAC name for the alcohol represented by this formula is _____.

32. The name for the compound

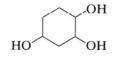

is _____.

     Whenever a compound contains more than one functional group, one of the groups is chosen as the principal group and is the basis of the name of the compound (that is, the suffix of the name). IUPAC rules include an order of priority for principal groups. (A partial list is in the Appendix.) All other functional groups outrank carbon-carbon multiple bonds.

     IUPAC substitutive names of unsaturated alcohols are formed by using <u>alkenol</u> or <u>alkynol</u> rather than alkanol as the basis of the name. The position of the OH group is assigned the lower possible number. In cycloalkenols, the OH group is always on the number 1 carbon. No locant is necessary for it in the name, but the locant 1 is usually included anyway. The locant for the carbon-carbon multiple bond may precede the parent stem of the name (traditional) or the <u>en</u> ending (new), and the locant for the OH group immediately precedes <u>ol</u>. For example,

$$CH_3 - CH = CH - CH - CH_3$$
$$|$$
$$OH$$

is named 3-penten-2-ol or pent-3-en-2-ol.

33. The structural formula for allyl alcohol is

_____ ,

and the substitutive name for allyl alcohol is _____.

34. A structural formula for cyclohex-2-en-1-ol is

_____ .

35. A substitutive name for

is _____.

36. Oblivon, a hypnotic, is 3-methylpent-1-yn-3-ol, which may be represented by the formula

_____ .

37. The chain of carbon atoms that will serve as the basis of the IUPAC substitutive name of

$$CH_3 - \underset{\underset{\displaystyle CH_2 - CH_3}{\overset{\displaystyle CH_3}{|}}}{C} = CH - \underset{}{CH} - \underset{\overset{\displaystyle |}{OH}}{CH} - CH_2 - CH_3$$

contains _____ carbons, and the portion of the name used to designate that chain together with
         (number)

the functional group is _____ . When the carbon chain is properly numbered, the OH group

will be on carbon numbered _____ , and the alkene linkage will be assigned position number _____ . The

parent name then becomes _____ . There are _____ substituents (not including the OH
                                                          (number)

group) on the parent chain. They are named _____ and _____ and are located on carbons

numbered _____ and _____ , respectively. The complete IUPAC substitutive name for the unsaturated

alcohol becomes _____ .

38. A structural formula for 5-chloro-3-phenylhex-3-en-2-ol is

_____ .

39. A traditional substitutive name for $CH_3 - \underset{\overset{\displaystyle |}{OH}}{CH} - CH = CH - \underset{\overset{\displaystyle |}{OH}}{CH} - \underset{\overset{\displaystyle |}{CH_3}}{CH} - CH_3$

is _____ .

40. A traditional substitutive name for $CH_3 - CH_2 - \underset{\overset{\displaystyle |}{CH_3}}{CH} -$ OH is

_____ .

Alkenols, like other alkenes, may exist as cis,trans isomers (see Chapter 3, items 37 through 52).
The prefix cis- or trans- [or (E)- or (Z)-] precedes the parent name fragment to which the geometri-
cal tag applies.

41. A structural formula for 2-methyl-<u>trans</u>-3-penten-2-ol is

_____.

42. A completely descriptive name, including identification of configuration (spatial arrangement), for the compound represented by the formula

$$HO-CH_2-CH-C=C-CH_2-CH-CH_2-CH_2-CH_2-CH_3$$
$$\qquad\qquad\quad |\qquad |\quad |\qquad\qquad |$$
$$\qquad\qquad\ CH_3\ \ H\ \ H\qquad\qquad C_6H_5$$

is _____.

43. By use of a geometric figure for the ring structure, 2-<u>cis</u>,6-<u>trans</u>-cyclodeca-2,6-dien-1-ol may be represented by the structure

_____.

44. Even so complex an alcohol as vitamin A,

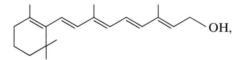

can be properly named rather easily by a substitutive name. The cyclic and acyclic portions of the line drawing shown have the same significance; that is, at each intersection or termination point, there is a carbon and as many hydrogens as necessary to complete a valence of 4. The ring portion of the compound

is a substituent on the last carbon of a parent chain of _____ carbons. That substituent is traditionally
(number)

named _____.

Each of the alkene linkages in the parent chain is _____. The full name of the substituted, unsaturated
(cis or trans)

alcohol illustrated is _____.

OH AS SUBSTITUENT

If an OH group is not attached to the parent chain (or is for other reasons to be treated as a substituent rather than the basis of the name), the prefix <u>hydroxy</u> is used to signify an OH substituent. This substituent name is used just as any other substituent name.

45. The compound whose substitutive name is 2-(hydroxymethyl)propane-1,3-diol has the formula

_____ .

46. The formula

$$CH_3$$
$$|$$
$$CH_2$$
$$|$$ $$OH$$
$$|$$
$$HO \!-\!\!\bigcirc\!\!-\!CH\!-\!CH_2\!-\!CH_2\!-\!CH\!-\!CH_3$$

shows on position 5 a substituent named _____ ; the name of

the compound is _____ .

ALKYL GROUPS REVISITED: NEW STYLE NAMES

Alkyl groups may now be named by the pattern you have learned for alcohols: The suffix yl signifies loss of a hydrogen from a parent hydride and replaces the final e of the hydride name. CH_4 is methane, and $CH_3 \!-\!$ is, by the new style nomenclature, methanyl. Just as ol can be associated with different locants, so can yl in this style nomenclature. $CH_3 \!-\! CH_2 \!-\! CH_2 \!-\!$ is propan-1-yl, and $CH_3 \!-\! CH \!-\! CH_3$ is propan-2-yl. The yl suffix out-ranks all other suffixes for lowest possible locant.
$\qquad\qquad\quad |$

Note that propyl (traditional style) does not include a locant, because the position of free valence is always 1, and the traditional name for $CH_3 \!-\! CH \!-\! CH_3$ is 1-methylethyl. With the publication in 1993
$\qquad\qquad\qquad\qquad\qquad\qquad\qquad\quad |$

of *A Guide to IUPAC Nomenclature of Organic Compounds*, both styles, 1-methylethyl and propan-2-yl, are correct and equally authoritative. The simplicity of alkanyl names will probably override their initial unfamiliarity and lead chemists to favor them for branched structures.

2-Propyl (or prop-2-yl) is not correct and has never been sanctioned by IUPAC. The distinction between stem + yl and alkan(e) + yl may seem subtle, but it is important.

47. Butan-2-yl is now a permitted name for the substituent group represented by the formula

_____ . Two other names are also approved, _____
                                                                      (traditional substitutive name)

and _____ .
        (trivial name)

48. The traditional name for the substituent represented by the formula

$$CH_3 \!-\! CH \!-\! CH \!-\! CH \!-\! CH_2 \!-\! CH_3$$
$$\quad\;\; | \quad\;\; | \quad\;\; |$$
$$\quad\;\; CH_3 \;\; CH_3$$

is _____ . The new rules permit the alternative name _____ .

49. The new name for the group $(CH_3)_3C$ 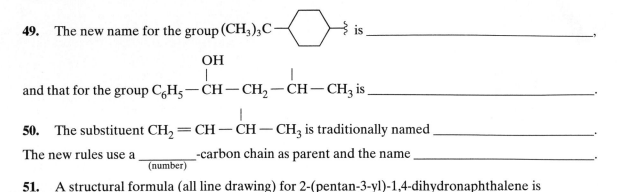 is _____,

and that for the group $C_6H_5 - \overset{\overset{\displaystyle OH}{|}}{CH} - CH_2 - \overset{|}{CH} - CH_3$ is _____.

50. The substituent $CH_2 = CH - \overset{|}{CH} - CH_3$ is traditionally named _____.

The new rules use a _____-carbon chain as parent and the name _____.
$\quad\quad\quad\quad\quad\quad$ (number)

51. A structural formula (all line drawing) for 2-(pentan-3-yl)-1,4-dihydronaphthalene is

_____.

Because the yl suffix has priority for low locant, the new rules for locants do not make a difference for cycloalkanyl and cycloalkenyl groups.

52. Cyclohexa-1,3-dien-1-yl is a name for the group represented by the formula

_____.

7

Ethers

Compounds having the structure represented by the general formula R — O — R' are classified as ethers. R and R' may be the same group (symmetrical ethers) or different groups (unsymmetrical ethers). Substitutive names are preferred for ethers, but multiple-word, functional class names are still used, especially for some familiar, symmetrical ethers. *Chemical Abstracts* does not use the functional class names any more in indexes.

Substitutive Names

There is no systematic suffix for the ether linkage in substitutive names. Ethers are named by considering R — O — as a substituent that replaces H in the parent compound, H — R'. The name of the R — O — substituent is formed by combining the name for the group R with <u>oxy</u>; for example, CH_3 — CH_2 — CH_2 — CH_2 — CH_2 — O — is pentyloxy or pentan-1-yloxy. A shortened name, illustrated by the general name <u>alkoxy</u>, is used when R is an alkyl group containing fewer than five carbons or is phenyl; for example, CH_3 — CH_2 — O — is ethoxy and C_6H_5 — O — is phenoxy. No particular preference is given to an oxy substituent over other substituents in assigning locants on the parent chain. Methoxymethane is the substitutive name for the symmetrical ether, CH_3 — O — CH_3.

1. The substitutive name of the ether CH_3 — CH_2 — O — CH_2 — CH_3 is _____.

2. The longest continuous chain of carbons in the formula

$$CH_3 — CH_2 — CH — CH_2 — CH — CH_3$$
$$CH_3 — CH_2 — O \qquad CH_3 — CH — CH_3$$

contains _____ carbons. The name for this chain is _____. Substituents
        (number)

appear on carbons numbered _____. Two of the substituents are called methyl, and the third is

called _____. A complete substitutive name for this ether is

_____.

3. A structural formula for 1,4-dimethoxybenzene is

_____,

and one for 1,1-diethoxybutane is

_____.

4. The substitutive name for the ether C_6H_5—O—C_6H_5 is _____.

5. The name of

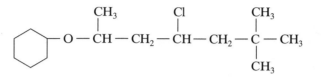

will be based on the parent compound, _____. There are _____ substituents on this

parent chain at positions numbered _____.

The complete name for the ether is _____.

6. A structural formula for 1,4-dimethoxy-<u>trans</u>-pent-2-ene is

_____.

7. A substitutive name for the alcohol

$$CH_3 — CH_2 — O \underset{}{\overset{OH}{\bigcirc}} — C_6H_5$$

will use _____ as the parent name and will specify two substituents, named _____

and _____. Correct numbering will assign the OH group to position number _____ and

the ethoxy substituent to position number _____. The complete substitutive name for this alcohol is

_____.

8. A substitutive name for the alkyne

$$CH_3 — C≡C — CH_2 — CH — O — CH — CH_3$$
$$| |$$
$$CH_2 — CH_2 — CH_2 — CH_2 — CH_3 CH_3$$

will be based on the longest continuous chain of carbons, containing _____ carbons.

The alkyne name for this chain is _____. The functional group that is the basis of the name is assigned position number _____, and the substituent, named _____, is on position number _____. The complete substitutive name for this compound is _____.

FUNCTIONAL CLASS NAMES

Ether is a class name, not the name of any individual member of the class. Names ending in ether are therefore multiple-word names, not single-word names, and are formed by preceding ether with the name of each group attached to oxygen, in alphabetical order. For example $CH_3-O-CH_2-CH_3$ may be named ethyl methyl ether, and CH_3-O-CH_3, dimethyl ether. Although IUPAC accepts functional class names for unsymmetrical ethers, some recommendations limit these names to symmetrical ethers.

9. Diphenyl ether may be represented by the formula

_____ ,

and isopropyl phenyl ether by the formula

_____ .

10. The compound represented by the formula $CH_2=CH-CH_2-O-CH_2-CH=CH_2$ shows attached to oxygen two groups that have the IUPAC common (trivial) name _____, and the compound represented by the formula has the functional class name _____.

11. Di-sec-butyl ether may be named with the substitutive name (either traditional or new),

_____.

12. The trivial name for the group

$$CH_3-CH_2-\underset{\underset{\displaystyle CH_3}{|}}{\overset{\overset{\displaystyle CH_3}{|}}{C}}-$$

is _____, the name of the group $\triangleright\!\!-$ is _____, and the functional class name for the compound

$$CH_3-CH_2-\underset{\underset{\displaystyle CH_3}{|}}{\overset{\overset{\displaystyle CH_3}{|}}{C}}-O-\triangleleft$$

is _____. The preferred, substitutive name for this compound

is _____.

POLYETHERS AND REPLACEMENT NOMENCLATURE

Compounds containing more than two ether linkages as part of a continuous chain of atoms are most easily named by a different style of IUPAC nomenclature, replacement names. The continuous chain is considered a parent chain, and the oxygen atoms are considered to be replacements for carbons in that chain. A multiplying prefix (di, tri, etc.) and the prefix oxa signify replacement of carbons in the parent chain by oxygens, and locants preceding the prefixes designate the location along the chain of those replacing atoms. For example, $CH_3 - O - CH_2 - CH_2 - O - CH_2 - CH_2 - O - CH_3$, a useful solvent nicknamed <u>diglyme</u>, may be properly named 2,5,8-trioxanonane. This replacement name, which tells us that oxygens replace carbons at positions 2, 5, and 8 in a nonane (nine-carbon) chain, is more convenient and is shorter than the alternative, substitutive name 1-methoxy-2-(2-methoxyethoxy)ethane. Although replacement names may be used for compounds with only one or two ether linkages, they offer little or no advantage over the substitutive names, which are favored for these compounds. Replacement names are intended to be used for compounds difficult to name in other approved ways.

Note that the prefix oxy signifies a substituent on a parent chain, and the prefix oxa signifies a replacement of a carbon in the parent chain.

13. 1,3,5-Trioxacyclohexane may be represented by the formula

_____.

14. The polyether $CH_3 - CH - O - CH_2 - CH - O - CH_2 - CH - O - CH_3$ may be conve-

$\qquad$ CH_3 $\qquad\qquad$ CH_3 $\qquad\qquad$ CH_3

niently considered as a parent chain of _____ atoms to which are attached_____ substituents.
$\qquad\qquad\qquad$ (number) $\qquad\qquad\qquad\qquad\qquad$ (number)

The replacing atoms take precedence over the substituents in numbering the parent chain. The re-

placement name for this compound is _____.

Replacement names may also be convenient for polyethers that contain other functional groups for which a systematic substitutive suffix is used.

15. 3,6-Dioxa-1,8-octanediol may be represented by the formula

_____.

16. The polyether $CH_2 = CH - O - CH_2 - O - CH_2 - CH_2 - O - CH_3$ may be named by the

replacement name _____.

Atoms replacing carbons in a parent compound are called <u>hetero atoms</u>, and replacement names are often convenient for compounds that illustrate multiple replacements of carbon by hetero atoms other than oxygen. A different prefix (ending in <u>a</u>) is used for each different replacing atom (<u>aza</u> for N, <u>thia</u> for S, <u>sila</u> for Si), and the citation sequence for those prefixes, rather than being strictly alphabetical, always begins with oxa (if present). (See the Appendix for priority sequences for these prefixes.) For example, $CH_3-O-CH_2-CH_2-NH-CH_2-CH_2-O-CH_3$ may be considered as a chain of nine atoms (with three hetero atoms in the chain) and named 2,8-dioxa-5-azanonane. The use of replacement names is likely to grow with use of computer-based literature searches, because the names are convenient for indexing.

17. A convenient, replacement name for the compound

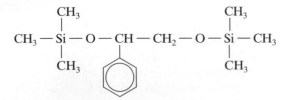

will be based on a parent chain of _____ atoms, including _____ hetero atoms. The oxygens are
(number)   (number)

cited first as replacing hetero atoms, and the other replacing atoms are designated by the prefix _____.

The name for the compound will be _____.

18. Eicosane is the IUPAC name for $C_{20}H_{42}$ (unbranched). 3,8,13,18-Tetraazaeicosane, which is effective in the treatment of AIDS-related diarrhea, may be represented by the formula

_____.

8

Substitution Products from Aromatic Hydrocarbons

Just as with aliphatic hydrocarbons, hydrogen in aromatic hydrocarbons may be replaced by a variety of substituent groups. Substituted benzenes (arenes) are named in essentially the same way as are substituted alkanes. For example, C_6H_5 — Cl is named chlorobenzene. Some other common substituents, in both alkanes and arenes, are: —F, fluoro; —Br, bromo; —I, iodo; —NO_2, nitro; —NO, nitroso; and —CN, cyano.

1. The substituted benzene represented by the formula

is named _____; that represented by the formula

is named _____; that represented by the formula

is named _____; and that represented by the formula

is named _____.

The relative positions of substituents in multiply substituted benzenes are indicated by letters (<u>o</u>-, <u>m</u>-, or <u>p</u>- for disubstituted compounds) or by numbers. Numerical locants are preferred. When numbers are used in the name of a substituted benzene, any substituent may be assigned position number 1, with the limitation that the set of lowest possible locants for substituents must be used.

2. Formulas for three isomeric dichlorobenzenes may be drawn

_____, _____, and _____;
the compounds represented by these formulas may be named, with letters used to designate relative
portions of substitution, _____, _____,
and _____, respectively, or, with numerical locants, _____,
_____, and _____, respectively.

3. <u>m</u>-Chloronitrobenzene may be represented by the formula

_____.

If numbers are used in the name, <u>m</u>-chloronitrobenzene may be named _____.

4. Picryl chloride is a common name for the very reactive compound

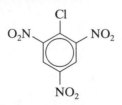

The lowest numerical locants that can be used in a systematic name for picryl chloride are _____,
_____, _____, and _____. A systematic name for this substituted benzene, with numerical locants
being used to indicate positions of substitution, is _____.

5. 2,4,6-Trinitrotoluene, commonly known as TNT, may be represented by the formula

_____.

6. Even though the formulas for picryl chloride (item 4) and TNT (item 5) closely resemble each
other, the locants for the nitro substituents are different when the compounds are named as derivatives
of different hydrocarbons. Different locants are used because _____
_____.

TNT (item 5) may also be named as a substituted benzene; its name then is _____
_____. Compare the locants in this name with those in item 4.

In compounds related to alkylbenzenes, a substituent may be bound to a carbon in the benzene nucleus or to a carbon in the alkyl group. The alkyl group is called a <u>side chain</u>.

7. Formulas may be drawn to represent three isomeric compounds formed by replacing a hydrogen in the aromatic nucleus of toluene by a chlorine. The three formulas are

_____ , _____ , and _____ .

and the three substituted toluenes may be named _____ , _____ ,

and _____ , respectively.

8. A fourth isomer, in which the chloro substituent appears on the side chain rather than in the benzene nucleus, may be represented by the formula

_____ .

When the substituent appears on the side chain of toluene, the entire side chain should be named as a substituent on the parent, benzene.

9. The compound 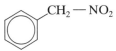 $CH_2 - Cl$ is named as a substituted benzene. The single substituent is

named chloromethyl (named toward point of attachment), and the substitutive name for the compound

is _____ .

10. The compound represented by the formula

$$CH_2 - NO_2$$

is named _____ .

Parentheses around the names of the substituents are needed in items 9 and 10 because _____

_____

_____ .

For purposes of nomenclature, the group

$$- CH_2 -$$

is often regarded as an alkyl group and is named <u>benzyl</u>. The name benzyl is restricted to the unsubstituted group and to the parent group with substituents in the ring only.

11. A functional class name for the compound represented by the formula

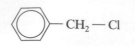

is _____; a functional class name for the compound

CH₂—I (structure)

is _____; and a functional class name for the compound

CH₂—O—CH₂ (structure)

is _____.

12. Benzyl alcohol, represented by the formula

_____ , is classified as a _____ alcohol.
                                                                                  (primary, etc.)

Benzyl is accepted by IUPAC, but *Chemical Abstracts* uses phenylmethyl instead of benzyl for this group in its indexes.

Phenyl and benzyl are names that are sometimes confused by beginning students.

13. Phenylmagnesium bromide is a Grignard reagent represented by the formula

_____,

and benzylmagnesium bromide is represented by the formula

_____.

14. Benzyl bromide,

_____,
          (formula)

and p-bromotoluene,

_____,
(formula)

are isomers.

15. Recall that styrene may serve as a parent name for ring-substituted styrenes. Styrene, an important industrial chemical, is represented by the formula

_____.

16. 4-Nitrostyrene may be represented by the formula

_____; it will be indexed in

Chemical Abstracts as a disubstituted benzene, _____.
(compound name)

17. 2,4-Difluorostyrene may be represented by the formula

_____ .

18. If the compound represented by the formula

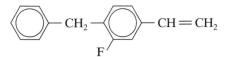

is named as a substituted styrene, the substituent in the 4-position will be named _____, and

the name of the compound will be _____.

19. Naphthalene is represented by the formula

_____,

and 1,5-dibromonaphthalene is represented by the formula

_____.

20.

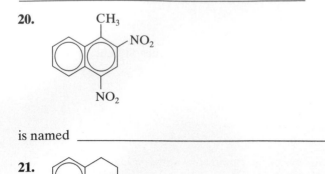

is named _____.

21.

may be named as a naphthalene; on that basis, hydro will be used to designate _____

_____,

and the name of the compound will be _____.

22.   Biphenyl is represented by the formula

_____,

and 2,4'-dibromo-4-nitrosobiphenyl is represented by the formula

_____.

9
Acids

CARBOXYLIC ACIDS

$$\overset{O}{\overset{\|}{-C}} - OH$$

The functional group $-\overset{\overset{O}{\|}}{C} - OH$ (or $-COOH$ or $-CO_2H$) is called a <u>carboxy group</u>, and compounds containing this functional group are carboxylic acids. Several types of IUPAC-approved names are used for carboxylic acids; structural features of the compound other than the carboxy group often determine the type of name most frequently used.

SUBSTITUTIVE NAMES (ACYCLIC CARBOXYLIC ACIDS)

If one imagines that a terminal CH_3 group of an acyclic hydrocarbon (alkane, alkene, or alkyne) is transformed into a COOH group, the acid is named by combining the hydrocarbon name (minus the final <u>e</u>) with the systematic suffix, <u>oic acid</u>. The stem of the acid name indicates the number of carbons in the parent compound, including the one in the carboxy group. For example, $CH_3 - CH_2 - COOH$ is propanoic acid.

In such names, the C in the COOH group is always position number 1 (lowest locant for suffix), but the locant is not included in the name. This locant assignment takes precedence over substituents and other functional groups in -oic acid names.

1. The structural formula for butanoic acid is _____, and that

for but-2-enoic acid is _____.

2. The substitutive name for the acid

$$CH_3 - CH - CH_2 - COOH$$
$$\quad\quad\quad | $$
$$\quad\quad CH_2 - CH_2 - CH_3$$

will be based on a parent chain of _____ carbons, for which the stem is _____.
    (number)

A substituent named _____ is at position number _____. The name of the acid

is _____.

3. The compound $CH_2 = CH - CH_2 - CH_2 - CH_2 - CH_2 - CH_2 - CH_2 - CH_2 - COOH$ is named _____.

4. The parent chain in the formula

$$CH_3 - CH_2 - \underset{\underset{CH_3}{|}}{CH} - CH_2 - \underset{\underset{NO_2}{|}}{\overset{\overset{C_6H_5 - CH_2}{|}}{C}} - CH_2 - COOH$$

contains _____ carbons. Substituents appear on carbons numbered _____.
         (number)

The name for this acid is _____.

5. Substituents named ethynyl and ethoxy are represented by the formulas _____ and _____, respectively. A structural formula for 5-ethoxy-2-ethynylpentanoic acid is

_____ .

6. Naphthalene has the formula

_____ ,

and 2-naphthyloxy is a substituent group with the formula

_____ .

2-(2-Naphthyloxy)ethanoic acid has the formula

_____ .

7. $C_6H_5 - C \equiv C - CH_2 - CH_2 - COOH$ will be named _____.

8. The parent compound on which the name of

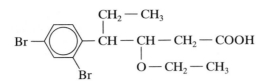

will be based is named _____, and the

substituents are named _____.

The name of the compound illustrated is _____.

9. Because of the geometry of some multiple-bond linkages, cis,trans isomers are possible for

_____ acids but not for _____ acids.
<u>(alkenoic or alkynoic)</u>   <u>(alkenoic or alkynoic)</u>

10. In the name 2-ethyl-<u>trans</u>-pent-3-enoic acid, <u>trans</u> refers to _____

_____, and a structural formula for the compound is

_____. A structural formula for the isomer, 2-ethyl-<u>cis</u>-pent-3-enoic acid, is

_____.

Acyclic compounds containing two carboxy groups are named by combining the ending -<u>dioic acid</u> with the name of the corresponding parent hydrocarbon. For example, HOOC — COOH may be named ethanedioic acid. Note that the treatment of the final <u>e</u> of the hydrocarbon name parallels that in the substitutive names for alcohols: The <u>e</u> is dropped for addition of a vowel ending (-ol and -oic acid) but is retained for addition of a consonant ending (-diol and -dioic acid).

11. Hexanedioic acid contains _____ carbons and is represented by the formula
       <u>(number)</u>

_____.

12. The compound

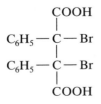

may be named _____.

13. The compound

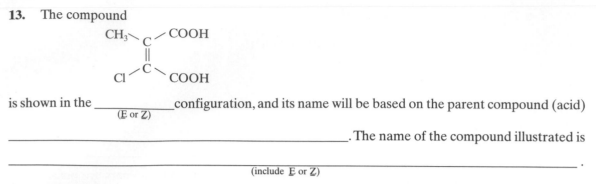

is shown in the _____ configuration, and its name will be based on the parent compound (acid)
              (E or Z)

_____. The name of the compound illustrated is

_____ .
                            (include E or Z)

14. (E)-3,7-Dimethyloct-2-enedioic acid is a component of a bean weevil copulation release hormone,

nicknamed <u>erectin</u>; the structural formula for this compound is _____ .

SUBSTITUTIVE NAMES (CYCLIC CARBOXYLIC ACIDS)

Compounds in which a carboxy group is attached to a ring system are named by combining the name of the ring system with the suffix -<u>carboxylic acid</u>. For example, ▷—COOH is cyclopropanecarboxylic acid. Note that the stem of the name does <u>not</u> include the carbon in the carboxy group ("<u>carb</u>oxylic" counts a carbon) and that the final <u>e</u> is retained for combination with a consonant suffix. Except in rings with fixed numbering (naphthalene, for example), the carbon to which the carboxy group is attached is the number 1 position, and all other locants follow from that one. The locant 1 does not usually appear in the name. Names of compounds containing more than one carboxy group include the appropriate multiplier in the ending (for example, -dicarboxylic acid), and locants for all carboxy groups must be used for clarity.

15. 4-<u>tert</u>-Butylcyclohexanecarboxylic acid may be represented by the formula

_____ .

16. The isomers

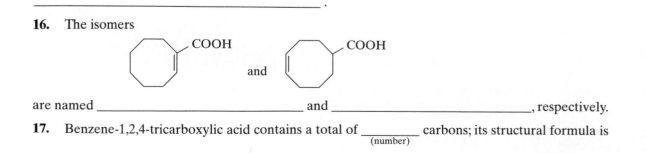

and

are named _____ and _____ , respectively.

17. Benzene-1,2,4-tricarboxylic acid contains a total of _____ carbons; its structural formula is
                                                            (number)

_____ .

18. The naphthalene derivative,

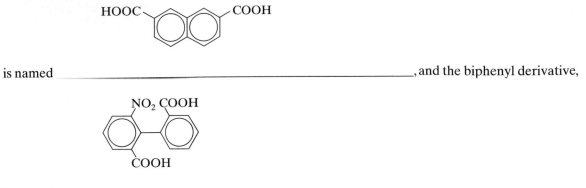

is named _____, and the biphenyl derivative,

is named _____.

19. 1,4-Dihydronaphthalene-2-carboxylic acid is represented by the formula

_____ .

Acyclic compounds containing more than two carboxy groups are conveniently named in this style, since it permits the maximum number of the same functional group to be treated alike.

20. The name propane-1,1,3,3,-tetracarboxylic acid indicates a total of _____ carbons in the compound. The compound has the formula
$\underset{\text{(number)}}{}$

_____ .

21. The compound

$$\text{HOOC} - \text{CH}_2 - \underset{\underset{\text{COOH}}{|}}{\text{CH}} - \text{CH}_2 - \underset{\underset{\text{COOH}}{|}}{\text{CH}} - \text{CH}_2 - \text{COOH}$$

may be named with a name that treats all the carboxy groups alike; that name is _____
_____. When all the carboxy groups <u>cannot</u> be treated alike in the name, some (the minimum number) are treated as substituents for which the prefix carboxy is used.

TRIVIAL NAMES (ARENECARBOXYLIC ACIDS)

Some arenecarboxylic acids are nearly always identified by shortened, trivial names.

22.   Benzenecarboxylic acid, whose formula is

_____ ,

is usually named benzoic acid, and the isomeric naphthalenecarboxylic acids,

_____ and _____ ,
                (formula)   (formula)

are usually named naphthoic acids. IUPAC accepts these trivial names, as well as the longer, more systematic ones, and the trivial ones are used by most chemists. *Chemical Abstracts* uses benzoic acid for that simple acid but uses the longer, -carboxylic acid names for some benzoic acid derivatives and for other arenecarboxylic acids. Locants for substituents are used in the same way in both kinds of names.

23.   3,5-Dinitrobenzoic acid may be represented by the formula

_____ .

24.   The formula

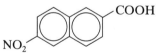

shows a carboxy group attached to the _____ position of naphthalene. The compound may be named
                                                              (number)

by the longer IUPAC name used by *Chemical Abstracts* index, _____ ,

or by the shorter IUPAC trivial name used by many chemists, _____ .

25.   The name of the compound

$$CH_3 - C \equiv C - \bigcirc - COOH$$

may be based on a parent acid with the (IUPAC) trivial name _____ .

The substituent is named _____ , and the compound illustrated is named

_____ .

TRIVIAL NAMES (ALKANOIC ACIDS)

Unsystematic names of some carboxylic acids were so well established by the time that rigid rules for systematic nomenclature were attempted by an International Congress in 1892 that they could not be discarded. These trivial names have been retained for some low-molecular-weight carboxylic acids (those containing fewer than six carbons) and are approved by the IUPAC rules for the unsubstituted acids (exception: acetic acid may be used even with substitution). *Chemical Abstracts* uses the trivial names only for the first two alkanoic acids, however.

The trivial names of alkanoic acids are formed by adding to the proper stem the ending ic and the separate word acid. The stems used for carboxylic acids are quite different from those learned for alkyl groups. They are usually Latin or Greek in origin and often reflect the natural sources from which the acids were first isolated. The stems are associated with particular numbers of carbons and structures just as the stems for alkyl groups are.

Trivial names for some alkanoic acids are listed below. The stems, which will figure in names of acid derivatives, are underlined.

formic acid	$H - COOH$
acetic acid	$CH_3 - COOH$
propionic acid	$CH_3 - CH_2 - COOH$
butyric acid	$CH_3 - CH_2 - CH_2 - COOH$
isobutyric acid	$CH_3 - \underset{\underset{CH_3}{\mid}}{CH} - COOH$
valeric acid	$CH_3 - CH_2 - CH_2 - CH_2 - COOH$
isovaleric acid	$CH_3 - \underset{\underset{CH_3}{\mid}}{CH} - CH_2 - COOH$

Only the stems of names signifying three and four carbons in carboxylic acids resemble the stems of names of alkyl groups. Locants for substituents are used in these trivial names as usual; the C of the carboxy group is position number 1.

26. The stem for three carbons in an alkyl group is _____, and that for three carbons in a carboxylic acid is _____. The stem for four carbons in an alkyl group is _____, and that for four carbons in a carboxylic acid is _____.

All other stems for the trivial names of carboxylic acids are completely different from the stems used for alkyl groups of corresponding carbon content.

27. The most common carboxylic acid contains two carbons and may be represented by the structural formula _____. The trivial name for this acid is _____.

28. The stem acet will always signify _____ carbons in a carboxylic acid or acid derivative.
(number)

29. Since there is only one position available for a substituent in acetic acid, no locant is necessary.

The trivial name for the substituted acid represented by the formula $Cl—CH_2—COOH$ is

_____, and the trivial name for the acid represented by the formula

$Cl_3C—COOH$ is _____.

Note that chloroacetic and trichloroacetic are single words. Acetic acid is the name of an individual compound, and modifiers (substituent prefixes) must be written as part of the same word.

30. <u>tert</u>-Butoxy is the name of the substituent represented by the formula

_____ ,

and <u>tert</u>-butoxyacetic acid may be represented by the formula

_____ .

31. Methanoic acid is the systematic name for the compound represented by the formula

_____ but the trivial name, _____ is actually used more frequently than the systematic name for this compound.

PEROXY ACIDS

The $—O—O—$ linkage is a peroxide linkage, and acids that contain $—O—OH$ in place of $—OH$ are called <u>peroxy acids</u>. (Note that peroxy acid is a two-word designation.) Peroxy acids are named by incorporating the unseparated prefix <u>peroxy</u> into the name of the corresponding acid. For acids named with <u>-oic acid</u> or <u>-ic acid</u> suffix, peroxy precedes the stem portion of the name; for example $CH_3—\overset{\displaystyle ||}{\underset{\displaystyle O}{C}}—O—OH$ is peroxyacetic acid. (For convenience, a peroxy acid may be represented by a formula such as $CH_3—CO_3H$.) For acids named with the <u>-carboxylic acid</u> suffix, <u>peroxy</u> immediately precedes carboxylic.

32. The peroxy acid of lowest molecular weight is HCO_3H, which is named

_____.

33. <u>m</u>-Chloroperoxybenzoic acid has the formula

_____ .

34. Cyclohexaneperoxycarboxylic acid has the formula

_____ .

35. The compound $CF_3 - CO_3H$ is named _____.

    Although a few frequently used peroxy acids have been identified by shortened names in which <u>per</u> replaces <u>peroxy</u> as the prefix, discontinuance of such names is recommended. Consistent use of the prefix peroxy makes naming (and comprehension) easier.

SULFONIC ACIDS

Carboxylic acids are organic acids that contain a carboxy group, but other organic acids include the <u>sulfonic acids</u>, compounds that contain a sulfo group, $- SO_3H$, bound to carbon. Sulfonic acids are named in the same style as are cyclic carboxylic acids; that is, the suffix <u>sulfonic acid</u> is combined with the name of the parent hydrocarbon (final <u>e</u> retained), and the sulfo group is given the lowest possible locant. For example, $CH_3 - CH_2 - CH - SO_3H$ is butane-2-sulfonic acid.
$$\qquad\qquad\qquad\qquad\qquad\qquad\qquad\qquad\quad |$$
$$\qquad\qquad\qquad\qquad\qquad\qquad\qquad\quad CH_3$$

36. Benzenesulfonic acid has the formula _____, and 4-methylbenzene-1-sulfonic acid has the formula

_____ .

4-Methylbenzene-1-sulfonic acid is often called <u>p</u>-toluenesulfonic acid; that is, toluene rather than benzene is considered the parent hydrocarbon. *Chemical Abstracts* uses the benzene name, however, which is simpler in approach though longer, and IUPAC is likely soon to limit use of toluene as a parent hydrocarbon to compounds for which substituents can be cited only as prefixes.

37. The name of the compound

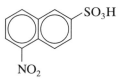

will be based on the parent hydrocarbon, _____. The sulfo group will

have the lowest possible locant, _____, and the compound is named

_____ .

38. The compound $CF_3 - SO_3H$ is named _____.

39. $\begin{matrix} CH_3 \\ \diagdown \\ \diagup \end{matrix}$ — SO_3H is named _____.

10
Acid Derivatives

ACID ANHYDRIDES

Removal of an OH group from a carboxylic acid generates an acyl group ($R-\overset{\overset{\displaystyle O}{\|}}{C}-$ or $R-CO-$).

If two acyl groups are attached to oxygen ($R-\overset{\overset{\displaystyle O}{\|}}{C}-O-\overset{\overset{\displaystyle O}{\|}}{C}-R$, or for convenience in typing, $R-CO-O-CO-R$), the compound is an <u>acid anhydride</u>. An acid anhydride is related to an acid by the loss of HOH between two molecules of acid. If R and R' are the same, the compound is a symmetrical acid anhydride; if different, an unsymmetrical acid anhydride.

Symmetrical acid anhydrides are named simply by replacing <u>acid</u> in the name of the corresponding acid with <u>anhydride</u>.

1. The trival name of the most common carboxylic acid, CH_3-COOH, is _____,

and the name of the related compound, $CH_3-CO-O-CO-CH_3$, is _____.

2. Benzoic acid has the formula _____, and benzoic anhydride

has the formula _____.

3. Trifluoroacetic anhydride has the formula _____.

4. The acid

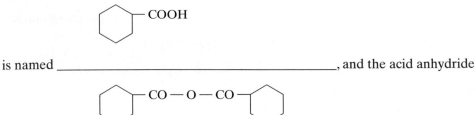

is named _____, and the acid anhydride

is named _____.

The same type of nomenclature is used for acid anhydrides related to acids other than carboxylic acids.

5. The strong acid $CF_3 — SO_3H$ is named _____, and the

acid anhydride $CF_3 — SO_2 — O — SO_2 — CF_3$ is named _____.

    Cyclic anhydrides of dicarboxylic acids are well known and are named in the same way as are acyclic anhydrides.

6. Benzene-1,2-dicarboxylic acid has the formula

_____,

and the cyclic anhydride related to it, benzene-1,2-dicarboxylic anhydride, has the formula

_____.

7. Butanedioic anhydride may be represented by the formula

_____ .This compound is usually identified by its trivial name, succinic anhydride.

8. The acid anhydride represented by the formula

is related to the dicarboxylic acid named _____,

and the anhydride itself is named _____.

    Unsymmetrical acid anhydrides are named by using the name of each related acid (except for <u>acid</u>) as a separate word (alphabetically arranged) followed by the separate word <u>anhydride</u>. That is, the name is three words.

9. The acid anhydride represented by the formula $C_6H_5 — CO — O — CO — CH_3$ is related to

two different carboxylic acids, commonly called _____ and

_____ .The name of the acid anhydride is _____.

10. The unsymmetrical acid anhydride

$$\langle\bigcirc\rangle — SO_2 — O — CO — CH_2 — CH_2 — CH_3$$

is named _____.

ACID HALIDES

Compounds in which the OH group in an acid functional group is replaced by a halogen are called <u>acid halides</u> or <u>acyl halides</u>. They are named by two-word, functional class names. The group attached to halogen is named by modifying the name of the corresponding acid: For an acyclic acid name, the suffix <u>-ic acid</u> is changed to <u>yl</u> (for example, methanoic acid to methanoyl); for a cyclic acid name, the suffix <u>carboxylic acid</u> is changed to <u>carbonyl</u>.

11. The acid halide C_6H_5—CO—Cl is related to the acid named _____

and is itself named _____.

12. Acetic acid has the formula _____, and acetyl chloride has the formula

_____.

13. The acid represented by the formula

$$CH_3 — CH_2 — CH — COOH$$
$$|$$
$$Br$$

is named _____, and the compound

$$CH_3 — CH_2 — CH — CO — Br$$
$$|$$
$$Br$$

is named _____.

14. 4-Nitrobenzenesulfonyl chloride has the formula

_____.

15. The compound ▷— CO — Cl is named _____.

16. The compound

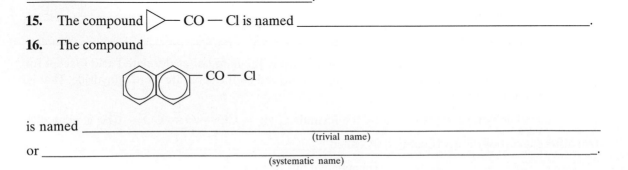

is named _____
(trivial name)
or _____.
(systematic name)

ESTERS

Replacement of an OH group in an acid by an OR group (R = alkyl or aryl) gives an <u>ester</u>. If the ester is related to a carboxylic acid, the ester may be represented by the general formulas

$$\overset{\overset{\displaystyle O}{\underset{\displaystyle \|}{}}}{R' - C - O - R} \quad \text{or} \quad R' - COOR \quad \text{or} \quad R - O - CO - R'$$

Note that the R group is always clearly attached to O and the R' group to $C=O$ in these illustrations. Esters may be related to acids other than carboxylic acids; for example, to nitric acid (acid, $HO-NO_2$; ester, $RO-NO_2$).

Esters are named in the same manner as salts (even though esters and salts are completely different from each other in properties): Two-word names are used. The R group (alkyl or aryl) is named as the first word, and the second word is formed by modifying the name of the acid to which the ester is related. The ending <u>ic acid</u> is replaced by <u>ate</u>. For example, acet<u>ic acid</u> (ethan<u>oic acid</u>) and nit<u>ric acid</u> will each form an ester, which will be called acet<u>ate</u> (ethan<u>oate</u>) and nit<u>rate</u>, respectively.

17.   Methanoic acid is more often called by its trivial name, _____.

The second word of the systematic name of an ester related to this acid will be _____,

and the second word of a trivial name for the ester will be _____.

18.   Propionic acid is the trivial name for the acid whose systematic name is _____.

The second word of the trivial name of an ester related to this acid will be _____, and

of a systematic name, _____.

19.   An ester formed from butyric acid will be called a _____.

20.   Methyl butyrate has the formula _____.

21.   The formula

$$\overset{\overset{\displaystyle O}{\underset{\displaystyle \|}{}}}{\underset{}{}} \qquad \overset{\displaystyle CH_3}{\underset{\displaystyle |}{}}$$
$$\bigcirc\!\!-\!C - O - CH - CH_2 - CH_3$$

shows attached to oxygen an alkyl group named _____. The name of the

alkyl group is the first word in the name of the ester. The acid to which the ester is related is named

_____, and the second word in the name of the ester

becomes _____. The complete two-word name for the

ester is _____.

22.   A structural formula for trifluoracetic acid is

_____ ,

a structural formula for the alkyl group isobutyl is

_____ ,

and a structural formula for isobutyl trifluoroacetate is

_____ .

23. Isobutyric acid may be represented by the formula

_____ ,

and isobutyl isobutyrate by the formula

_____ .

24. A trivial name for

$$CH_2 = CH - CH_2 - O - \overset{\displaystyle O}{\overset{\displaystyle \|}{C}} - H$$

is _____ .

25. Phenyl propionate may be represented by the structural formula

_____ ,

and its isomer, benzyl acetate, by the formula

_____ .

26. To name a complex ester such as

$$CH_3 - CH = C - CH_2 - CH - CH_2 - COO - CH_2 - CH - CH_2 - CH - CH_3$$

with the Cl and CH_3 substituents on the acyl side and $O-CH_3$ and CH_3 on the alkyl side.

we consider the alkyl group and the acyl group separately. The alkyl group of the ester is in the

_____ half of the formula shown, and the basis of its name will be a chain of _____ carbons
(right or left)   (number)

with substituents on positions numbered _____ and _____. The substituents are named

_____ and _____, respectively, and the name of the entire

alkyl group is _____.

The acyl group contains a parent chain of _____ carbons with a chloro substituent on position
                                          (number)

number _____ and a methyl substituent on position number _____. The complete name of the

carboxylic acid to which the ester is related is _____,

and the last word of the two-word ester name will be _____.

The complete name of the ester will be _____.

27. The ester

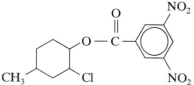

will have as the second word of its name _____.

The alkyl group will be named _____,

and the ester will be named _____.

28. A structural formula representing <u>trans</u>-pent-2-en-1-yl 4-methylbenzenesulfonate is

_____ .

29. The compound represented by the formula

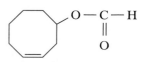

may be named _____.

30. Boric acid may be represented by the formula $(HO)_3B$, and tributyl borate by the formula

_____ .

31. The ester

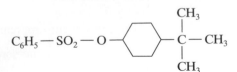

is related to an acid named _____ . The second word

of the two-word ester name is _____ , and the full

name of the ester is _____ .

32. Vinyl trifluoromethanesulfonate has the formula

_____ .

33. Dodecane is the IUPAC name for the unbranched alkane with 12 carbons. A widely used detergent,

sodium dodecyl sulfate, is both a salt and an ester related to the inorganic acid, _____ acid.
A structural formula for the compound (detergent) is

_____ .

34. (Z)-Dodec-7-en-1-yl acetate, emitted by both elephants and moths to signal readiness for mating,

may be represented by the structural formula _____ .

35. The ester represented by the formula

$$CH_3 - O - CH_2 - CH_2 - O - CH_2 - CH_2 - O - CO - CH_3$$

may be named readily with a replacement name. The replacement name of the alkyl group is

_____ ; the name of the ester is

_____ .

When an ester group is to be named as a substituent (rather than serve as the basis of the name),
it is named an <u>alkyloxycarbonyl</u> or <u>aryloxycarbonyl group</u> (sometimes shortened to alkoxycarbonyl

or phenoxycarbonyl; see Chapter 7). For example, $CH_3 — O — CO —$ as a substituent is called methoxycarbonyl. Note that a substituent with a combination name is always named <u>toward</u> the point of attachment, not <u>from</u> that point.

36. An acid group has higher priority than an ester group in naming (see Appendix). The name of the substituted acid

$$CH_3 — CH_2 — O — CO —\!\!\bigcirc\!\!— COOH$$

will include the substituent name, _____. The complete name

of the acid illustrated is _____.

37. 4-Phenoxycarbonyl-1-naphthoic acid is represented by the formula

_____ .

AMIDES

Compounds that contain an acyl group bonded to nitrogen are <u>amides</u>. Amides are named by replacing the ending in the name of the corresponding acid (<u>ic acid</u>, <u>oic acid</u>, or <u>ylic acid</u>, depending on the style of acid name used) by the systematic ending <u>amide</u>. For example, $CH_3 — \underset{\underset{O}{\|}}{C} — NH_2$ is related to

acetic acid (ethanoic acid) and is named acet<u>amide</u> (ethan<u>amide</u>).

38. The compound represented by the formula $H — \underset{\underset{O}{\|}}{\overset{\overset{O}{\|}}{C}} — NH_2$ is related to the acid

_____, and is named _____.
       (trivial name)

39. Benzamide is represented by the formula

_____ ,

and benzenesulfonamide by the formula

_____ .

40. The formula

$$\langle \rangle - \overset{\overset{\text{O}}{\|}}{C} - NH_2$$

represents a compound related to the acid named _____.

The amide is named _____.

Substituents may be on N as well as on positions in the acyl group. The locant <u>N</u> (underlined for italics) is used for substituents on N in exactly the same way as numerical locants are used for other substituents.

41. 4-Methylhexanamide is represented by the formula

_____,

and its isomer, <u>N</u>-methylhexanamide, is represented by the formula

_____.

42. <u>N,N</u>-Dimethylformamide is represented by the formula

_____.

43. The compound represented by the formula

$$\langle \bigcirc \rangle - SO_2 - NH - CH_2 - CH_2 - CH_2 - CH_3$$

is named _____.

44. The compound represented by the formula

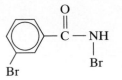

has two substituents on a parent amide named _____. The positions of the

substituents are designated by locants _____ and _____. The name of the substituted amide is

_____.

11

Aldehydes and Ketones

The group $-\overset{\overset{\displaystyle O}{\|}}{C}-$ is called a <u>carbonyl group</u>. Compounds that contain a $-\overset{\overset{\displaystyle O}{\|}}{C}-H(-CHO)$ group attached to hydrogen or carbon are classified as <u>aldehydes</u>. Compounds that contain a carbonyl group attached to two carbons are classified as <u>ketones</u>. These two classes are related to each other in much the same way as primary and secondary alcohols are related to each other. Although a single systematic suffix (ol) is used for the different classes of alcohols, chemists continue to use different systematic suffixes for aldehydes and ketones. Some nomenclature problems would be eased if a single suffix were used for these carbonyl-containing compounds, and such practice for general use has been considered.

In aldehydes, at least one hydrogen is attached to the carbonyl group; in ketones, two hydrocarbon groups (which may be substituted) are attached to the carbonyl group.

1. The type of compound represented by the formula

$$CH_3 - CH_2 - \overset{\overset{\displaystyle }{\underset{\underset{\displaystyle O}{\|}}{C}}}{} - CH_3$$

is _____.

2. The type of compound represented by the formula

$$CH_3 - \overset{\overset{\displaystyle O}{\|}}{C} - H$$

is _____.

3. The formula

$$H - \overset{\overset{\displaystyle }{\underset{\underset{\displaystyle O}{\|}}{C}}}{} - H$$

represents _____, and
<u>(type of compound)</u>

represents _____.
             (type of compound)

For convenience in typing and writing, aldehydes are often represented by a condensed structural formula such as R — CHO. Note that the functional group is written — CHO rather than —COH, to avoid any confusion with alcohols. Ketones may be represented conveniently by a formula such as R — CO — R.

SYSTEMATIC (SUBSTITUTIVE) NAMES

Ketones and acyclic aldehydes are named by replacing the final e of the name of the corresponding hydrocarbon (including the carbon in the carbonyl group) with the systematic suffix one (pronounced as in tone) or al (pronounced as in pal), respectively. For example, CH_3 — CHO is ethanal, and CH_3 — CO — CH_3 is propanone.

4.   The formula CH_3 — CH_2 — CH_2 — CH_2 — CHO represents an aldehyde containing five carbons.

The systematic name of the corresponding hydrocarbon is _____, and the systematic

name of the aldehyde is _____.

5.   Cyclohexanone may be represented by the formula

_____ .

6.   Propanal may be represented by the formula

_____ .

Since the aldehyde functional group must necessarily be at the end of the chain, it will always be position number 1 in parent chains named as aldehydes. No number designation for the aldehyde functional group is necessary in the name. The carbonyl group in ketones, on the other hand, is not restricted to one position, and a locant is required in the name if isomeric positions are possible. When the carbonyl-containing compound is named as an alkanone, the ketone functional group is given the lower possible locant in the parent chain containing it. In traditional IUPAC names, this locant precedes the parent chain stem (2-butanone); in the new style IUPAC names, it immediately precedes the suffix -one (butan-2-one). Substituent positions in both aldehydes and ketones are given the lower possible locants after the proper locant has been assigned to the carbonyl group.

7.   There are two isomers that may be called pentanone; structural formulas for these isomers may

be written _____ and _____.

The complete systematic names for these isomers are _____ or _____ and
                                                              (traditional)   (new)
_____ or _____, respectively.
     (traditional)   (new)

8. A structural formula for 3-chloro-2-pentanone may be drawn

_____ .

3-Chloropentanal may be represented by the formula

_____ .

9. The compound whose formula is

$$CH_3 - CO - \overset{\displaystyle CH_3}{\underset{\displaystyle CH_3}{\overset{|}{\underset{|}{C}}}} - CH_3$$

is named _____ .

10. The hydrocarbon corresponding to the aldehyde

$$CH_3 - \underset{\underset{\displaystyle C_6H_5}{|}}{CH} - CH_2 - CH = CH - CHO$$

will have the substitutive name _____, and the systematic name for the aldehyde itself

will be _____ .

11. The hydrocarbon corresponding to the aldehyde

$$CH_3 - \underset{\underset{\displaystyle C_6H_5}{|}}{C} = CH - CH_2 - CH_2 - CH_2 - CHO$$

will actually have the substitutive name _____. When the compound represented

by the formula is to be named as an aldehyde, however, the — CHO functional group takes precedence

over any other molecular fragment for numbering. The carbonyl group becomes position number _____,

the alkene linkage is assigned position number _____, and the phenyl substituent appears on position

number _____. The substitutive name for the aldehyde is _____ .

12. Two alarm substances produced by ants are the compounds represented by formulas A and B.

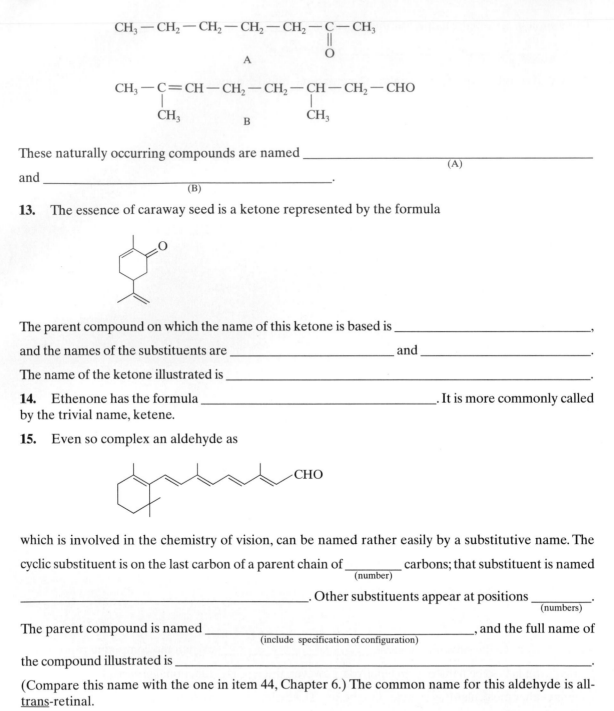

These naturally occurring compounds are named _____
                                                                                    (A)

and _____.
                              (B)

13. The essence of caraway seed is a ketone represented by the formula

The parent compound on which the name of this ketone is based is _____,

and the names of the substituents are _____ and _____.

The name of the ketone illustrated is _____.

14. Ethenone has the formula _____. It is more commonly called by the trivial name, ketene.

15. Even so complex an aldehyde as

which is involved in the chemistry of vision, can be named rather easily by a substitutive name. The

cyclic substituent is on the last carbon of a parent chain of _____ carbons; that substituent is named
                                                              (number)

_____. Other substituents appear at positions _____.
                                                                                      (numbers)

The parent compound is named _____, and the full name of
                                    (include specification of configuration)

the compound illustrated is _____.

(Compare this name with the one in item 44, Chapter 6.) The common name for this aldehyde is all-<u>trans</u>-retinal.

16. The compound

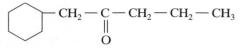

is named as a substituted alkanone. The substituent, on position number _____, is named

_____. The traditional name of the ketone illustrated is _____.

The name of the isomer of the foregoing ketone,

is 1-cyclohexyl-1-pentanone, even though the parent compound, 1-pentanone, is actually an aldehyde and is named pentanal. The <u>one</u> ending is the approved one for all similar compounds and clearly conveys the correct functional group information. This use illustrates the appeal of a single systematic suffix for aldehydes and ketones.

17.   1,4-Diphenyl-1-hexanone is represented by the formula

_____ .

18.   The ketone

is named _____ .

Compounds containing two like carbonyl groups are named by combining the ending <u>dial</u> (two syllables) or <u>dione</u> with the name of the corresponding parent hydrocarbon. Locants are used for the carbonyl groups in names of diones but are not used in names of dials (the two CHO groups are the termini of the parent chain).

19.   Hexanedial is represented by the formula _____, and

hexane-2,4-dione by the formula _____ .

20. C_6H_5 — $CH(CH_2$ — CH_2 — $CHO)_2$ is a condensed formula representation of the aldehyde named _____ .

CYCLIC ALDEHYDES

When one or more — CHO groups are attached directly to a cyclic system, the compound is named by adding the suffix <u>carbaldehyde</u>, <u>dicarbaldehyde</u>, etc., to the name of the cyclic system. Except for cyclic systems with fixed numbering (naphthalene, for example), the carbon to which a — CHO group is attached is the number 1 position, and a locant for the — CHO group in a monoaldehyde is not need-ed or used. For example, 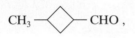 CHO is cyclopropanecarbaldehyde. Locants for all — CHO groups in a polyaldehyde are required; they precede the name of the cyclic system in traditional names or the suf-fix in new style names.

    Chemical Abstracts uses the suffix carboxaldehyde rather than carbaldehyde, but the IUPAC suffix avoids the redundancy of both <u>ox</u> and <u>aldehyde</u> indicating oxygen.

21. The name for

⬠— CHO

is _____, and the name for its isomer,

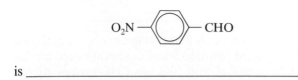

is _____.

22. Naphthalene-2,6-dicarbaldehyde is represented by the formula

_____ .

23. The systematic name for

O_2N —⬡— CHO

is _____.

TRIVIAL NAMES OF ALDEHYDES

Several simple aldehydes are commonly named by IUPAC-approved names based on the approved trivial names of the corresponding carboxylic acids (see Chapter 9). The ending <u>aldehyde</u> replaces <u>ic</u> <u>acid</u> or <u>oic acid</u> in the trivial name of the acid. For example, H — CHO is form<u>aldehyde</u>.

24. Acetic acid has the formula _____, and acetaldehyde has the formula

_____ .

25. The trivial name for the acid corresponding to the aldehyde

$$CH_3 - \overset{\overset{\displaystyle CH_3}{|}}{CH} - CHO$$

is _____, and the trivial name for the aldehyde is _____

_____.

26. Benzenecarbaldehyde is the systematic name for the aldehyde whose formula is _____.

The trivial name of the acid corresponding to this aldehyde is _____, and the trivial

name of the aldehyde itself is _____, which is nearly always used.

27. The trivial name of the aldehyde $CCl_3 - CHO$ is _____.

FUNCTIONAL CLASS NAMES OF KETONES

Some ketones are usually named by multiple-word names ending in <u>ketone</u>. The <u>ketone</u> portion of the name refers to the carbonyl group, and the two groups attached to the carbonyl group are named alphabetically and separately. If both groups are alike (symmetrical ketones), the multiplying prefix <u>di</u> is used with the name of the group.

28. The formula for ethyl methyl ketone is _____. The preferred name for this ketone is the systematic name, 2-butanone or butan-2-one.

29. The functional class name for 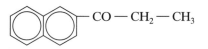 is _____,

and that for $C_6H_5 - CO - C_6H_5$ is _____.

Symmetrical ketones of this type (two cyclic groups) are named by *Chemical Abstracts* with systematic names based on methanone as the parent compound; that is, diphenylmethanone instead of diphenyl ketone. Note again that the unsubstituted parent compound in this example is actually an aldehyde.

30. The ketone

$$\text{[naphthalene ring]} - CO - CH_2 - CH_3$$

may be named with the functional class name, _____, or with the

systematic name, _____.

31. Benzyl <u>sec</u>-butyl ketone has the formula

_____

and the traditional systematic name _____.

TRIVIAL NAMES OF KETONES

A few ketones are commonly known by names containing the stems of the trivial names of related carboxylic acids. Acetone ($CH_3 — CO — CH_3$) is an example of this kind of name and is the only trivial name for an acyclic ketone whose continued use is approved. Trivial names continue to be used for some unsymmetrical ketones in which one of the groups is phenyl (phenyl <u>not</u> substituted). For these ketones, the ending <u>ophenone</u> replaces the <u>ic acid</u> or <u>oic acid</u> ending of the trivial name of the carboxylic acid corresponding to the remainder of the ketone (that is, the acyl group). For example, $CH_3 — CO — C_6H_5$ is commonly known as acetophenone. The shortened stem <u>propi</u> rather than <u>propion</u> for $CH_3 — CH_2 — CO —$ is used in this kind of name.

32.   The trivial name for $C_6H_5 — CO — C_6H_5$ is _____.

33.   2,2,2-Trifluoroacetophenone has the formula

_____, and hexadeuterioacetone (deuterio = ^2H or D replaces H) has the formula

_____ .

CARBONYL GROUPS AS SUBSTITUENTS

Some aldehydes and ketones contain other functional groups that may be chosen as the basis of the name of the compound. For example, a compound may contain both a ketone functional group and a carboxylic acid functional group. (The relative priorities of functional groups for naming multifunctional compounds are given in a table in the Appendix; in general, the more bonds from carbon to hetero atoms in the functional group, the higher the priority of that functional group for naming.) If another functional group has priority as the basis of the name, the oxygen of the aldehyde or ketone carbonyl group is treated as a substituent. The prefix <u>oxo</u>, signifying the $=O$ substituent (not the carbon also), is used just as any other substituent prefix; it does not have any special preference in assignment of locants. Note that the prefix for $=O$ as a substituent on a parent chain is ox<u>o</u>, while that for $—O—$ as a replacement for carbon in a parent chain is ox<u>a</u>.

34.   4-Oxocyclohexanecarboxylic acid has the formula

_____, and 4-oxacyclohexanecarboxylic acid has the formula

_____ .

35. The structural formula for 6-oxohexanoic acid may be drawn

_____ .

In addition to being a carboxylic acid, this compound may also be classified as a(n) _____ .

36. "Queen substance," a bee sex attractant isolated from queen honey bee glands, is 9-oxo-<u>trans</u>-dec-2-enoic acid, which may be represented by the structural formula

_____ .

37. The compound represented by the formula

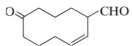

may be named as an aldehyde; the basis of the name (parent name), without specifying numbering or substituent, will be _____ , and proper numbering of the ring will assign position number _____ to the alkene linkage. The complete name for the substituted aldehyde will be _____ .

A — CHO group to be treated as a substituent is named formyl or methanoyl.

38. The compound represented by the formula

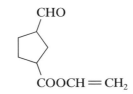

is named as a substituted ester. It is named _____ .

40. 4-Formylbenzoic acid may be represented by the formula

_____ .

12
Amines and Related Cations

Amines are organic compounds related to ammonia, both in structure and chemical behavior. The trivalent nitrogen atom in amines is bound only to hydrocarbon groups (which may be substituted) and hydrogens. If only one hydrogen of ammonia (amine) is replaced by a hydrocarbon group (for example, $CH_3 — NH_2$), the amine is classified as a primary amine; if two hydrocarbon groups are bound to the nitrogen, the amine is classified as a secondary amine; and if three hydrocarbon groups are bound to nitrogen (no hydrogen bound to nitrogen), the amine is a tertiary amine.

1. The compound represented by the formula $CH_3 — CH_2 — NH — CH_2 — CH_3$ is classified as

a _____ amine.

2. The compound represented by the formula

$$CH_3 — \underset{\underset{CH_3}{|}}{N} — CH_3$$

is classified as a _____ amine.

3. When hydrocarbon groups are represented by the general symbol, R, primary amines may be represented by the generalized formula _____, secondary amines by the generalized

formula _____, and tertiary amines by the generalized formula_____.

Note that the classification of amines depends on the degree of substitution on the nitrogen atom, not on the nature of the hydrocarbon group or groups. The classification of alcohols, on the other hand, depends on the nature of the hydrocarbon group, not on the oxygen, since only one group can be attached to oxygen in an alcohol.

4. The alcohol represented by the formula

$$CH_3 — \underset{\underset{CH_3}{|}}{\overset{\overset{CH_3}{|}}{C}} — OH$$

is classified as a _____ alcohol, and the amine represented by the formula

$$CH_3 - \overset{\overset{\displaystyle CH_3}{|}}{\underset{\underset{\displaystyle CH_3}{|}}{C}} - NH_2$$

is classified as a _____ amine.

5.   The alcohol represented by the formula

$$CH_3 - CH_2 - \overset{}{\underset{\underset{\displaystyle OH}{|}}{CH}} - CH_3$$

is classified as a _____ alcohol, and the amine represented by the formula

$$CH_3 - CH_2 - \overset{}{\underset{\underset{\displaystyle NH_2}{|}}{CH}} - CH_3$$

is classified as a _____ amine.

6.   The alcohol represented by the formula

$$CH_3 - \overset{}{\underset{\underset{\displaystyle CH_3}{|}}{CH}} - CH_2 - OH$$

is classified as a _____ alcohol, and the amine represented by the formula

$$CH_3 - \overset{}{\underset{\underset{\displaystyle CH_3}{|}}{CH}} - CH_2 - NH_2$$

is classified as a _____ amine.

Primary amines are named by one-word substitutive names formed in one of three ways: (1) On the basis of the parent hydrocarbon to which $-NH_2$ is attached. The final e of the name of the parent hydrocarbon is replaced by the systematic suffix amine; a locant included in the usual way specifies the position of attachment of $-NH_2$. For example, $CH_3 - CH_2 - CH_2 - NH_2$ is 1-propanamine (*Chemical Abstracts* uses this style) or propan-1-amine. These names parallel the substitutive names of alcohols (for example, 1-propanol or propan-1-ol); the amine functional group takes precedence over substituents for numbering the parent chain. (2) On the basis of the parent compound, amine (synonymous in this context with ammonia, NH_3). The name of the hydrocarbon group attached to $-NH_2$ modifies amine in a one-word name. Amine is the name of an individual compound, NH_3. No locant is needed or used, because the hydrocarbon group can be attached only to N. For example, $CH_3 - CH_2 - CH_2 - NH_2$ is propylamine. *Chemical Abstracts* does not use this style. (3) As in (2), except that the systematic name azane is used instead of amine for NH_3.

The first style (based on parent hydrocarbon) is recommended for all primary amines, but the second style is more generally used for amines of simple structure.

Symmetrical secondary and tertiary amines with simple groups (all alike) attached to N are named by style (2) above with the inclusion of the appropriate multiplying prefix (di, tri). For example, $(CH_3)_3N$ is trimethylamine.

7. The amine represented by the formula

$$CH_3 - CH_2 - \underset{\underset{CH_3}{|}}{CH} - NH_2$$

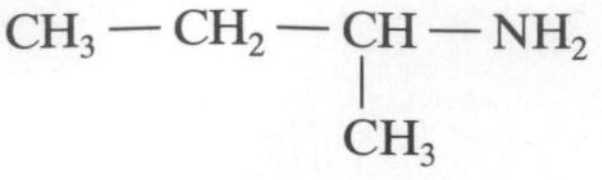

may be named by three names: one based on parent hydrocarbon, _____,

one based on parent amine, _____, and one based on

parent azane, _____. The larger parent is generally preferred.

8. <u>tert</u>-Butylamine may be represented by the formula

_____.

9. $C_6H_5 - NH - C_6H_5$ may be named _____.

10. Heptan-3-amine is the preferred name for the compound represented by the formula

_____.

11. The compound

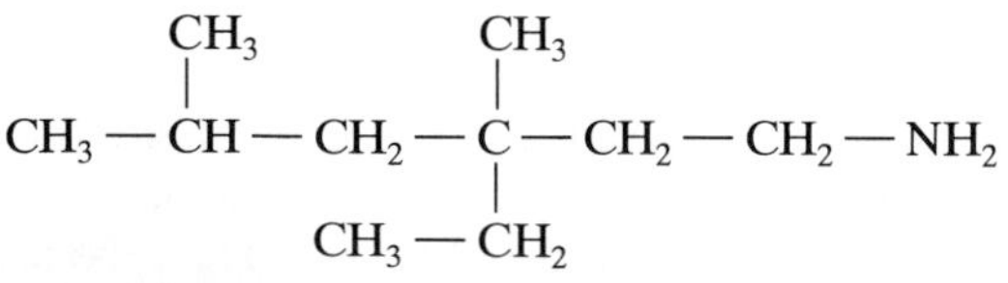

$$CH_3 - \underset{\underset{CH_3 - CH_2}{|}}{\underset{|}{CH}} - CH_2 - \underset{\underset{CH_3}{|}}{\overset{\overset{CH_3}{|}}{C}} - CH_2 - CH_2 - NH_2$$

is preferably named on the basis of the parent hydrocarbon. The name is _____

_____.

12. The compound

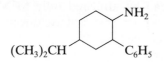

may be named on the basis of a parent hydrocarbon. The name is _____

_____.

13. Tricyclopropylamine may be represented by the formula

_____.

Unsymmetrical secondary and tertiary amines (all hydrocarbon groups not alike) are named as N-substituted derivatives of a primary amine. The primary amine with the largest or most complex hydrocarbon group is chosen as the parent amine for the name, and the other groups on N are treated as substituents with locant $\underline{N}$ (capital; underlined for italics). For example, $CH_3 — CH_2 — NH — CH_3$ is named $\underline{N}$-methylethanamine. IUPAC rules permit the name $\underline{N}$-ethylmethylamine and $\underline{N}$-ethylmethylazane, but the parent hydrocarbon-type name is preferred, because it uses the larger parent.

14. The amine represented by the formula

$$CH_3 — \underset{\underset{CH_3}{|}}{CH} — CH_2 — NH — \underset{\underset{CH_3}{|}}{CH} — CH_3$$

may be named as a derivative of the primary amine, _____.

The unsymmetrical amine may be named _____.

15. The parent hydrocarbon on which a systematic name of $C_6H_5 — N(CH_3)_2$ will be based is named

_____, and the tertiary amine $C_6H_5 — N(CH_3)_2$ may be named

_____. This compound is more frequently identified by its IUPAC-approved trivial name, $\underline{N},\underline{N}$-dimethylaniline, but *Chemical Abstracts* uses the systematic, parent hydrocarbon name in indexes.

16. $\underline{N},\underline{N}$-Dimethyl-4-vinylcyclohexanamine may be represented by the formula

_____. The name used here is based on _____

as parent compound.

17. $\underline{N}$-Phenyl-4-nitronaphthalen-1-amine is represented by the formula

_____.

Note that the name 3-pentanamine is correct, because the parent, pentane, has a position 3 to which the functional group, amine, can be attached, whereas the name 3-pentylamine is incorrect, because 3-pentyl is not a correct name for the alkyl substituent on the parent, amine.

The <u>amine</u> functional group is low in the priority order for citation as a suffix (see the Appendix for order of priority), but when <u>amine</u> is the ending of an IUPAC name, the $—NH_2$ group (amino group) takes precedence over other substituents or functional groups for numbering the parent chain.

18. The preferred name (based on parent hydrocarbon) for

$$CH_3 - CH - CH_2 - CH_2 - CH - CH_3$$
$$\quad\quad\quad | \quad\quad\quad\quad\quad\quad\quad\quad | $$
$$\quad\quad\quad CH_3 \quad\quad\quad\quad\quad\quad\quad NH_2$$

will indicate the amine functional group on carbon number _____. A complete name for the amine

will be _____.

19. The parent hydrocarbon name (without locant) on which the name of the amine

$$O - C_6H_5$$
$$|$$
$$CH_3 - CH_2 - CH - CH = CH - CH_2 - CH_2 - CH_2 - N(CH_3)_2$$

is based is _____, and the name of the amine itself

is _____.

20. A name, based on hydrocarbon parent, for the amine represented by the formula

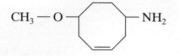

is _____.

21. The formula

$$CH_3 - CH_2 - CH - CH = C - CH_3$$
$$\quad\quad\quad\quad | \quad\quad\quad\quad | $$
$$\quad\quad\quad\quad CH_3 \quad\quad N - CH_2 - CH_3$$
$$\quad\quad\quad\quad\quad\quad\quad\quad | $$
$$\quad\quad\quad\quad\quad\quad\quad\quad CH_3$$

indicates a substituted amine that is preferably named on the parent hydrocarbon basis. Locants for

the methyl substituents are _____, and the locant for the ethyl substituent is _____. The

complete name for this substituted amine is _____.

In the names of compounds that contain other functional groups with higher priority as the basis of the name, the $-NH_2$ functional group is treated as a substituent named <u>amino</u>. In the IUPAC order of priority for name suffixes, amines are lower than nearly all other functional groups.

22. The compound represented by the formula

$$CH_3 - CH_2 - CH - CH - CH_2 - CH_3$$
$$\quad\quad\quad\quad | \quad\quad | $$
$$\quad\quad\quad\quad OH \quad NH_2$$

is named as an alcohol. The substituent will be called _____, and the full name

of the compound will be _____.

23. The compound represented by the formula

$$(CH_3)_2N - CH_2 - \underset{\underset{CH_3}{|}}{CH} - CH_2 - CO - CH_3$$

is named as a ketone with a methyl substituent on position number _____ and a

_____ substituent on position number _____. The complete

substitutive name for this ketone is _____.

24. A structural formula for ethyl 2-diethylamino-3-ethoxydecanoate may be drawn

_____.

  Compounds containing more than one amine functional group (amino group) attached to a parent hydrocarbon are named by use of the appropriate multiplying prefix with the suffix <u>amine</u>. (Note that the final <u>e</u> of the hydrocarbon name is retained as usual before a suffix beginning with a consonant.) These names parallel those for alkanediols. For example, $H_2N - CH_2 - CH_2 - CH_2 - NH_2$ is propane-1,3-diamine.

25. $H_2N - CH_2 - CH_2 - CH_2 - CH_2 - NH_2$ and $H_2N - CH_2 - CH_2 - CH_2 - CH_2 - CH_2 - NH_2$, products produced by decaying flesh, are sometimes called <u>putrescine</u> and <u>cadaverine</u>, respectively. IUPAC

names (more systematic and prosaic) for these amines are _____ and

_____, respectively.

26. Hexane-1,6-diamine, used in the manufacture of nylon, may be represented by the formula

_____.

27. 3-Methypentane-2,4-diamine may be represented by the formula

_____.

28. The triamine

$$H_2N - CH_2 - \underset{\underset{NH_2}{|}}{CH} - CH_2 - CH_2 - CH_2 - NH_2$$

is named _____.

29. N,N,N',N'-Tetramethylethane-1,2-diamine has the formula _____.
Ethane-1,2-diamine, whose formula is _____, is often called by
the bivalent name, ethylenediamine.

Ammonium Compounds

When ammonia is hydronated ($NH_3 + H^+ \longrightarrow NH^+_4$), the ion formed is called ammonium. (Hydron is the IUPAC designation for natural-abundance H^+; proton is to be restricted to the cation from the isotope ^1H.) An ion formed by hydronation of an amine may be named 1) as an alkyl (or aryl) substituted ammonium (example: methylammonium, $CH_3 — NH^+_3$) or 2) by adding the modifying suffix ium (= $+ H^+$) to the systematic name of the amine based on parent hydrocarbon (example: methanaminium, $CH_3 — NH^+_3$). Ammonium and aminium salts are named with two-word names: The cation and anion parts of the name are separate words. For example, $CH_3 — NH^+_3 Cl^-$ is methylammonium chloride or methanaminium chloride.

30. The amine represented by the formula $(CH_3 — CH_2)_2NH$ is usually named _____,
and the salt derived from it, $(CH_3 — CH_2)_2\overset{+}{N}H_2 Cl^-$, is named _____.

31. 2-Naphthylammonium perchlorate is represented by the formula

_____.

32. 4-tert-Butyl-N,N-dipropylcyclohexan-1-amine is converted by hydronation into the ion named

_____.

33. Pyridine is the name of an important aromatic amine with the nitrogen in the ring: .

(Nicotine is an alkyl-substituted pyridine.) When pyridine is hydronated, the cation formed is repre-

sented by the formula _____ and is named _____.

34. Hydroxylamine, $HO — NH_2$, a commonly used reagent for aldehydes and ketones, is usually marketed as the salt under the name hydroxylamine hydrochloride. This salt can be represented by the

formula _____ and more systematically named _____.

Quaternary ammonium ions have no H bound to N; other ammonium ions have at least one hydrogen bound to nitrogen. Quaternary ammonium ions and compounds are named by preceding ammonium by the names of the four groups covalently bound to N, alphabetically arranged, all in one word; the name of the anion follows as a separate word.

35. Benzyltrimethylammonium chloride can be represented by the formula

_____.

36. The amine $[CH_3-(CH_2)_6-CH_2]_2N-CH_2-CH_3$ is named _____

_____,

and the salt, $[CH_3-(CH_2)_6-CH_2]_2\overset{+}{N}(CH_2-CH_3)_2\,\overset{-}{I}$, is named _____

_____.

37. With R symbolizing an alkyl group, a quaternary ammonium hydroxide may be represented by the generalized formula _____.

38. The biologically important quaternary ammonium hydroxide $HO-CH_2-CH_2-\overset{+}{N}(CH_3)_3\overset{-}{O}H$ is commonly called choline. Its systematic name is _____.

Some other biologically important compounds containing an ammonium group are internal salts (zwitterions). They usually are identified by trivial, unsystematic names, but they can be named with one-word systematic names, which are much better for communicating structural information to persons who have not memorized the trivial name identifications. The systematic names are generated by citing the cationic substituent ammonio (underline{o} replaces underline{um}) as a prefix and the anionic component as a suffix (one-word names). For example, $H_3N-CH_2-CO\overset{-}{O}$ is ammonioacetate (or ammonioethanoate). Note that amino is the name of an uncharged $-NH_2$ substituent, and ammonio is the name of a cationic $-\overset{+}{N}H_3$ substituent.

39. The amino acid commonly known as alanine is actually a zwitterion systematically named 2-ammoniopropanoate. It has the formula

_____. Ammonium propanoate, on the other hand, has

the formula _____. These two compounds

_____ isomers.
<u>(are/are not)</u>

40. The amino acid leucine is 2-ammonio-4-methylpentanoate and has the formula

_____.

41. Threonine has the formula

$$CH_3 - CH - CH - COO\text{\textunderscore}$$
$$\quad\quad\quad | \quad\quad |$$
$$\quad\quad OH \quad NH_3$$
$$\quad\quad\quad\quad\quad +$$

and the systematic name _____.

41. The zwitterion commonly known as betaine (three syllables) has the formula $(CH_3)_3\overset{+}{N}-CH_2-CO\overset{-}{O}$ and the systematic name _____.

13
Bridged Ring Systems

Hydrocarbon systems that have two or more carbons common to two or more rings are called <u>bridged hydrocarbons</u>. A combination of a multiplying prefix (<u>bi</u> for two) and the prefix <u>cyclo</u> specifies the number of rings. For example, a bicycloalkane is a saturated hydrocarbon whose structure is two rings joined through two common carbon atoms. The carbon atoms at the junctures of the rings are called <u>bridgeheads</u>, and the bonds, atoms, or chains of atoms connecting the bridgeheads are called <u>bridges</u>.

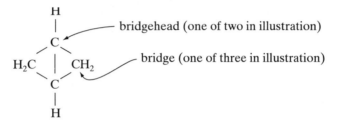

Although bridged ring systems containing more than two rings are well known, this chapter will be confined to bicyclo compounds.

The stem that replaces <u>alk</u> in the specific name of the bicycloalkane corresponds to the total number of carbons in the two rings. The formula above represents a bicyclobutane.

1. The hydrocarbon represented by the formula

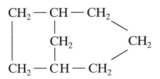

is a bicycloalkane containing a total of _____ carbons in the two rings; it will be called a
(number)

_____.

2. The hydrocarbon represented by the formula

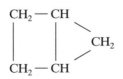

will be called a _____

3. The two bridgeheads in the bicyclooctane formula of item 1 are separated from each other by three bridges, containing _____, _____, and _____ carbons, respectively.
                                                   (number)   (number)   (number)

4. The two bridgeheads in the bicyclopentane formula of item 2 are separated from each other by three bridges, containing _____, _____, and _____ carbons, respectively.
                                                   (number)   (number)   (number)

These numbers, designating the length of the bridges, are used in the full name of a bicycloalkane to differentiate it from isomeric bicycloalkanes. The style is illustrated by the name bicyclo[3.2.1.]octane. Note that the numbers are arranged in descending order, are separated from each other by periods, and are enclosed in brackets. The name is a one-word name without any space separation between parts.

5. The full name for the bicyclopentane illustrated in item 2 is _____.

Bicycloalkanes, like cycloalkanes, are conveniently represented by geometric figures. The formulas

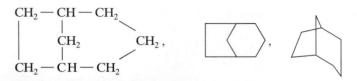

are equivalent representations of bicyclo[3.2.1]octane. The last figure is intended to show the actual geometry of the molecule.

Several isomeric bicyclooctanes may be represented by formulas, five of which are

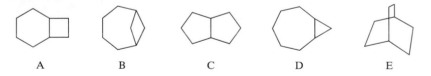

     A   B   C   D   E

These isomeric bicyclooctanes are differentiated from one another by full names that include numbers.

6. The distinguishing, full name for

    formula A is _____ ;

    formula B is _____ ;

    formula C is _____ ;

    formula D is _____ ;

    formula E is _____ .

Note that the numbers in brackets account for all carbons other than the bridgeheads in the bicycloalkane framework; the sum of these numbers always equals two less than the number of carbons signified by the stem in the name.

Numbering of a bicycloalkane to indicate location of substituents begins at one bridgehead, proceeds around the longest bridge to the other bridgehead, continues around the second longest bridge back to the number 1 position (original bridgehead), and is completed across the shortest bridge. The numbered formula for bicyclo[3.2.1.]octane illustrates the numbering.

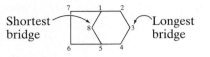

The choice of bridgehead for position number 1 is made to permit substituents to be assigned the lower possible position numbers.

7.   Correct numbering of the formula

will assign to the methyl substituent locant _____, whereas the methyl substituent in

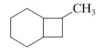

will be assigned locant _____.

8.   Correct numbering of the formula

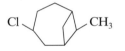

will assign to the methyl substituent locant _____ and to the chloro substituent locant _____.

9.   8,8-Dichlorobicyclo[5.1.0]octane may be represented by the formula

_____.

10.   1,3-Dimethylbicyclo[1.1.0]butane may be represented by the formula

_____.

As in naphthalene, the numbering of bicycloalkanes is fixed, and the locants for functional groups follow from that fixed numbering. Bridged hydrocarbons containing carbon-carbon double bonds in the ring system are named by replacing the final ending <u>ane</u> with <u>ene</u> and inserting a locant immediately before <u>ene</u> to indicate position of the alkene linkage. The bridgehead that will permit the alkene linkage to have the lower locant is chosen for position number 1.

11. The alkene represented by the formula

may be named with the systematic name _____.

12. α-Pinene (trivial name), the major constituent of turpentine, is 2,6,6-trimethylbicyclo[3.1.1]hept-2-ene; α-pinene may be represented by

_____.

13. Carene, an isomer of α-pinene, may be represented by the formula

The systematic name for carene is _____.

For bridged ring systems containing functional groups that will be the basis of the systematic name, the hydrocarbon name is followed by a locant and the appropriate systematic suffix for the functional group (final e of hydrocarbon name elided when suffix begins with a vowel).

14. The alcohol

may be named by the systematic name _____.

15. Camphor is a naturally occurring ketone represented by the formula

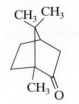

A systematic name for camphor is _____.

16. Bicyclo[4.4.0]decane-2-carboxylic acid may be represented by the formula

_____.

HETEROCYCLIC BRIDGED SYSTEMS

Some bridged ring systems containing ring atoms other than carbon are named conveniently by replacement names that include a prefix (<u>oxa</u> for O, <u>thia</u> for S, <u>aza</u> for N) to identify each replacing hetero atom.

17. 7-Oxabicyclo[4.1.0]heptane is the name by which *Chemical Abstracts* indexes the compound commonly called cyclohexene oxide and represented by the formula

_____ .

18. The compound represented by the formula

may be named with the systematic name _____ .

19. The name for the hydrocarbon represented by the formula

is _____; the prefix signifying replacement
of carbon by nitrogen is _____; and the systematic name for the compound represented by the formula

is _____ .

20. 1-Azabicyclo[2.2.2]octane (sometimes called <u>quinuclidine</u>) may be represented by the formula

_____ ,

and 7-thiabicyclo[2.2.1]heptane by the formula

_____ .

21. Cocaine, a local anesthetic that is now a major commodity in the illegal drug trade, has the structure

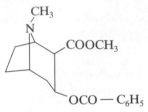

and is systematically named as an ester (priority order for functional groups).   The substituent
— O — CO — C$_6$H$_5$ is named benzoyloxy and is on position number _____. The parent acid (without substituents) is named _____, and the ester illustrated is named _____.

BICYCLO[2.2.1]HEPTANE

Derivatives of bicyclo[2.2.1]heptane have been cited so frequently in the chemical literature that some special attention to nomenclature of these derivatives is appropriate. The IUPAC-approved trivial name for 1,7,7-trimethylbicyclo[2.2.1]heptane is <u>bornane</u>. In polycyclic compounds of this class, IUPAC once approved exceptional use of <u>nor</u> as a prefix to indicate that <u>all</u> methyls attached to the ring skeleton of the parent compound were missing. Bicyclo[2.2.1]heptane was often named norbornane. This use of <u>nor</u> has been specifically abandoned by IUPAC, and <u>nor</u> is restricted to signifying loss of <u>one</u> carbon unit (CH$_2$ or CH$_3$).

22. Bornane is the trivial name for the compound represented by the formula

_____ , and 1,7,7-trinorbornane is the currently-approved trivial name for the compound represented by the formula

_____. Norbornane was a once-approved name for 1,7,7-trinorbornane, but it cannot now be used correctly for that compound. Because the trivial and systematic names are now virtually equally simple, there is little reason to use the trivial name anymore.

   The geometry of some substituted bicycloalkanes (such as bicyclo[2.2.1]heptane) is such that a substituent on the main ring may extend "within" or "outside" the obtuse angle of the main ring. The italicized prefixes <u>endo</u> ("within") and <u>exo</u> ("outside") are used to differentiate such isomers.

23. <u>exo</u>-Bicyclo[2.2.1]heptan-2-ol may be represented by the formula

and <u>endo</u>-bicyclo[2.2.1]heptan-2-ol by the formula

_____ .

24. is the formula for _____ .

25. may be named _____ .

26. <u>exo</u>-Bicyclo[2.2.1]heptan-2-yl acetate may be represented by the formula

_____ ,

and <u>endo</u>-bicyclo[2.2.1]heptan-2-yl benzenesulfonate by the formula

_____ .

14
Nomenclature of Reaction Intermediates

The transient nature of reaction intermediates does not reduce the need for precise nomenclature for them. The rules for systematic nomenclature of these neutral and charged species follow the same pattern you have learned for the more stable classes of compounds: A parent hydride name is modified by a suffix that signifies the particular type of functionality. Systematic suffixes for some reaction intermediates are given in the following table.

Parent Hydride	Formal Operation	Intermediate	Suffix
$R-H$	minus $H^{\cdot}$	$R^{\cdot}$	-yl
R_2CH_2	minus $2H^{\cdot}$	$R_2C:$	-ylidene
$R-H$	minus H^+	R^-	-ide
$R-H$	minus H^-	R^+	-ylium
$R-H$	plus H^+	RH_2^+	-ium

In traditional IUPAC names, these suffixes are always associated with locant 1 (except in structures with fixed numbering, such as bicycloalkanes). Example: 2-ethyl-1-propyl. In the new style IUPAC names, these suffixes may be associated with any position on the parent chain, just as the ol suffix for alcohols may be. Example: pentan-3-yl.

RADICALS

Neutral organic species that contain at least one unpaired electron are called <u>radicals</u>. (These species have frequently been called free radicals, but IUPAC has recommended that the "free" be dropped.) Structural formulas for radicals generally include a dot for the unpaired electron close to the symbol for the atom with that electron. Most radicals that are equivalent to an organic compound minus a hydrogen atom have specific names that end in <u>yl</u>. If a substituent group name ends in <u>yl</u>, that is the name of the corresponding radical; for example, $CH_3-\dot{C}H_2$ is ethyl. If a substituent group name ends in <u>o</u> or <u>y</u>, that terminal letter is changed to <u>yl</u> for the name of the corresponding radical; for example, $CH_3-\ddot{\underset{\cdot\cdot}{O}}\cdot$ is methoxyl. The name of a monovalent radical always ends in <u>yl</u>.

For clarity the separate word "radical" may be used with the group name: ethyl radical. IUPAC rules do not require "radical," and *Chemical Abstracts* does not use it for indexing, but, for complete clarity in general communication, inclusion of the class label may be useful.

Alkyl radicals are classified as primary, secondary, and tertiary in the same way as alcohols are.

1. The radical with the IUPAC trivial name <u>tert</u>-butyl is represented by the formula

_____ and is classified as a _____ radical.
                                                                    (primary, etc.)

2. Benzyl radical is represented by the formula _____ and is

classified as a _____ radical.
                (primary, etc.)

3. The new style IUPAC name for the radical $CH_3-CH_2-\dot{C}H_2$ is _____,

and that for $CH_3-CH_2-\dot{C}H-CH_2-CH_3$ is _____.

4. $(C_6H_5)_3C\cdot$ is named _____.

5.

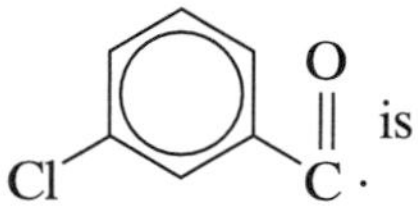

is named _____, and

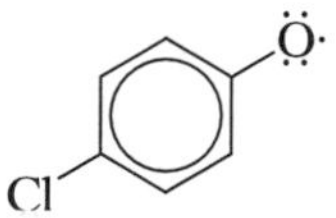

is named _____.

6. The benzoyloxyl radical has the formula _____, and the

hexan-3-aminyl radical has the formula _____.

7. The radical

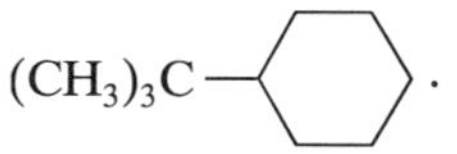

is named _____ and classified as a _____ radical.
                                                              (primary, etc.)

8. The radical

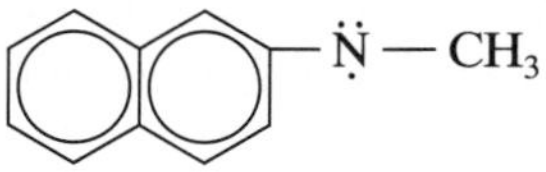

is named _____.

9. Cyclohexanecarbonyl radical may be represented by the formula _____, and

acetoxyl radical by the formula _____.

10. The common IUPAC name for $CH_3 - \overset{\overset{\displaystyle O}{\|}}{C} - NH_2$ is _____,

and that for $CH_3 - \overset{\overset{\displaystyle O}{\|}}{C} - \overset{\displaystyle .}{N}H$ is _____.

Radicals formally formed by loss of H from more than one carbon are named by use of the appropriate multiplying prefix (di, tri, etc.) with the suffix yl. Example: ethane-1,2-diyl.

11. The simplest way to name $\overset{\displaystyle .}{C}H_2 - \overset{\displaystyle .}{C}H - \overset{\displaystyle .}{C}H_2$ is by the new style IUPAC nomenclature. The name

is _____. The final e of the parent hydride name is _____
                                                                                                                                 (retained/elided)

because _____.

The radical at locant 1 is classified as a _____ radical, that at locant 2 as a _____ radical,
                                                                      (primary, etc.)   (primary, etc.)

and that at locant 3 as a _____ radical.
                                               (primary, etc.)

METHYLENES/ALKYLIDENES AND RELATED NITROGEN COMPOUNDS

Neutral particals that contain a divalent carbon (with six electrons) are classified as <u>carbenes</u> or <u>methylenes</u>. These structures may be named (1) on the basis of the parent compound, $\overset{\displaystyle ..}{C}H_2$, or (2) by use of a systematic suffix for a group with two free valences on the same carbon.

Carbene and methylene are synonymous names for the same parent compound, $\overset{\displaystyle ..}{C}H_2$, and both are used. The IUPAC preference is methylene, however, and *Chemical Abstracts* does not use carbene. Substituted methylenes (carbenes) are named in much the same way as substituted methanes are.

Formulas for methylenes (carbenes) usually include two dots on the appropriate C to represent the unshared (nonbonding) electrons.

12. The name of $\overset{\displaystyle ..}{C}Cl_2$ is _____.

13. Dicyanomethylene has the formula _____, and diphenylmethylene has the

formula _____.

Alkylmethylenes are named by combining the systematic suffix <u>ylidene</u> with the name of the parent hydride, including the divalent carbon. For example, $CH_3 - \overset{\displaystyle ..}{C}H$ is ethanylidene. The ylidene suffix may be associated with any locant (lowest possible). For example, $CH_3 - \overset{\displaystyle ..}{C} - CH_3$ is propan-2-ylidene. Alternatively, for simple structures, the suffix may be combined with the <u>stem</u> signifying the number of carbons in the parent chain, including the divalent carbon, which must be at position 1 except in structures with fixed numbering. For example, $CH_3 - \overset{\displaystyle ..}{C} - CH_3$ is 1-methylethylidene. (A functional class name for $CH_3 - CHCl_2$ is ethylidene dichloride.)

14. The particle represented by the formula $CH_3 - CH_2 - \overset{\displaystyle ..}{C}H$ is named _____,

and the one represented by the formula $CH_3 - CH_2 - CH - CH_2 - CH_2 - \overset{\displaystyle ..}{C}H$ is named

_____.

15. The name of $CH_3 — CH_2 — CH_2 — \ddot{C}H — CH_3$ is _____.

16. A formula for benzylidene is _____.

17. 4-<u>tert</u>-Butylcyclohexylidene may be represented by the formula _____.

18. The particle $CH_2 = CH$ 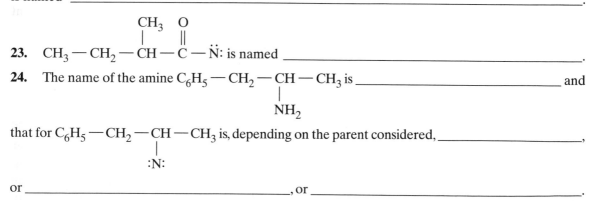 may be named _____.

19. Cyclopenta-2,4-dien-1-ylidene may be represented by the formula _____.

20. The name of the alkene 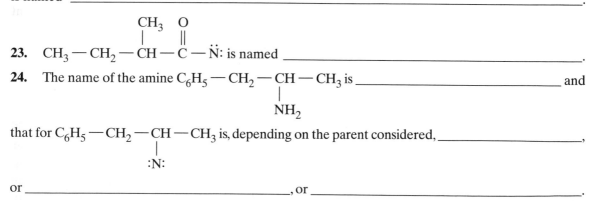 is _____;

The name of the methylene 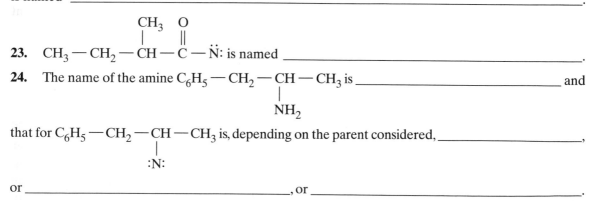 is _____;

Neutral particles that contain a monovalent nitrogen (with six electrons) are named on the basis of the parent compound <u>nitrene</u>, $H\ddot{\ddot{N}}$, or by combing the suffix ylidene with the parent name <u>azane</u> (for NH_3) or <u>alkanamine</u>.

21. Phenylnitrene is represented by the formula _____, and cyanonitrene by the formula _____.

22. A substituent with a combination name is named toward its point of attachment. The substituent $CH_3O — CO —$ is named _____, and the intermediate $CH_3 — O — CO — \ddot{N}$: is named _____.

23. $CH_3 — CH_2 — \overset{\overset{\displaystyle CH_3}{|}}{CH} — \overset{\overset{\displaystyle O}{||}}{C} — \ddot{N}$: is named _____.

24. The name of the amine $C_6H_5 — CH_2 — \underset{\underset{\displaystyle NH_2}{|}}{CH} — CH_3$ is _____ and

that for $C_6H_5 — CH_2 — \underset{\underset{\displaystyle :N:}{|}}{CH} — CH_3$ is, depending on the parent considered, _____,

or _____, or _____.

CARBOCATIONS AND CARBANIONS

Organic structures containing a carbon with a charge are classified as <u>carbocations</u> (+ charge) and <u>carbanions</u> (– charge). These ions can be formed, at least conceptually, by the addition or removal of hydrogen ions (H^+ or H^-) to/from parent hydrides. The systematic suffixes used connote hydrogen ion addition or removal:

Suffix	Connotation
ylium	removal of hydride (H^-)
ide	removal of hydron (H^+)
ium	addition of hydron (H^+)

Examples: $CH_3 - CH_3$ ethane $CH_3 - CH_2^+$ ethanylium (ethylium)

 $CH_3 - CH_4^+$ ethanium $CH_3 - CH_2^-$ ethanide

These suffixes have high priority for lowest locant.

Alternatively, and a bit simpler in many cases, these ions may be correctly named by combining, as a separate word, the class name <u>cation</u> or <u>anion</u> with the name of the corresponding radical. Examples: ethyl cation, propan-2-yl anion. However, *Chemical Abstracts* does not use this style; it does not include the class labels in the names of the ions. So much confusion has been associated with the use and misuse of "carbonium ion" in the names of specific carbocations that IUPAC has strongly recommended complete discontinuance of this term in specific names.

Carbocations and carbanions are classified as primary, secondary, and tertiary on the same basis as radicals (and alcohols) are.

25.   The carbocation $CH_3 - CH_2 - \overset{+}{C}H - CH_2 - CH_3$ is named _____

_____ and is classified as a _____
                                                                    (primary, etc.)

carbocation. The name 3-pentyl cation is incorrect because _____

_____.

26.   The 3-nitrophenylmethyl anion is represented by the formula

_____ and is classified as a _____ carbanion.
                                                        (primary, etc.)

27.   Allyl cation is represented by the formula _____, and vinyl cation by the formula

_____. A one-word, systematic name for allyl cation is _____.

28.   The ion , named with a one-word, systematic name, is _____

_____ and is classified as a _____ ion.
                                                            (primary, etc.)

29.   7-Methylbicyclo[2.2.1]hept-2-en-7-ylium has the formula

_____.

30.   The ion $(C_6H_5)_2\overset{..}{C} - CH_2 - C_6H_5$ is named _____

and is classified as a _____ ion.
(primary, etc.)

31.   The five species A, B, C, D, and E are named as follows:

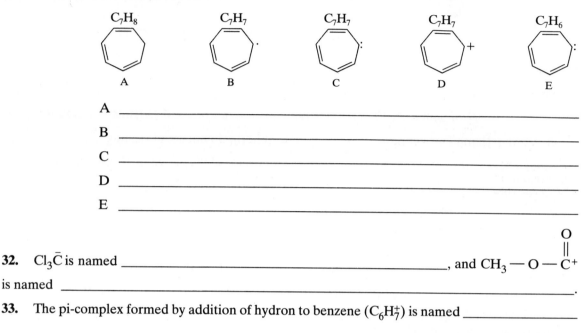

A _____

B _____

C _____

D _____

E _____

32. $Cl_3\overset{-}{C}$ is named _____, and $CH_3 - O - \overset{\overset{\displaystyle O}{\|}}{C}{}^{+}$

is named _____.

33.   The pi-complex formed by addition of hydron to benzene $(C_6H_7^+)$ is named _____

_____.

Appendix

- Selected References
- A Brief Summary of Some Key IUPAC Rules for Substitutive Names of Organic Compounds
- Substitutive Name Prefixes and Suffixes for Some Important Functional Groups
- Names of Some Important Parent Compounds Not Specifically Included in This Program
- Replacement Name Prefixes for Hetero Atoms

SELECTED REFERENCES

CAHN, R. S., and O. C. DERMER, *An Introduction to Chemical Nomenclature*, 5th ed., Butterworths, London, 1979.

Chemical Abstracts, 9th Collective Index, 1972–76, Index Guide.

FLETCHER, J. H., O. C. DERMER, and R. B. FOX, *Nomenclature of Organic Compounds* (*Advances in Chemistry Series* No. 126), American Chemical Society, Washington, D.C., 1974.

HURD, C. D., "The General Philosophy of Organic Nomenclature," *Journal of Chemical Education,* 38, 43 (1961).

IUPAC, *Nomenclature of Organic Chemistry, Sections A, B, C, D, E, F* and *H*, Pergamon Press, Oxford, 1979.

A Guide to IUPAC Nomenclature of Organic Compounds. Recommendations 1993. Prepared for pub. by R. Panico, W. H. Powell, K.-C. Richter. Blackwell Scientific Publications, London. 1993.

A Brief Summary of Some Key IUPAC Rules for Substitutive Names of Organic Compounds

1. The functional group to be used as the basis of the name must be identified. IUPAC rules include a full priority order of functional groups for names; a partial list is included in this Appendix. In general, a functional group with carbon in higher oxidation state (more bonds to hetero atoms) takes precedence over one with carbon in lower oxidation state. Radicals and ions are at the top of the priority order.

2. The longest chain of carbon atoms containing the functional group is the basis of the substitutive name. Atoms or groups other than hydrogen attached to the parent chain are called substituents.

3. The parent compound is named by adding the appropriate systematic ending (suffix) to the name of the corresponding hydrocarbon; the final e in the hydrocarbon name is elided (dropped) for a suffix beginning with a vowel but is retained for a suffix beginning with a consonant.

4. With a few exceptions, only the principal functional group is connoted by a systematic suffix; all others are treated as substituents. The exceptions are structures containing multiple carbon-carbon bonds (designated only by suffixes) and those containing a combination of radical and ion centers connoted by different suffixes.

5. The parent chain is numbered so that the functional group that is part of the parent compound is assigned the smaller possible number (locant). If numbering the parent chain in both directions gives the same locant to the functional group, the parent chain is numbered so that the set of lower locants is used for the substituents. The set of lower locants has the lower locant at the first point of difference when alternative sets in sequence are compared term by term.

6. Locants for functional group(s) in the parent compound immediately precede the suffix (new IUPAC style) or precede the hydrocarbon (stem) portion of the name (traditional IUPAC style). When the systematic ending clearly requires that the functional group include the terminal carbon of the parent chain (for example, -al for aldehyde), and when the suffix is associated with position 1 in a ring, the locant 1 may be omitted from the name.

7. A locant for each substituent must appear in the name even when the same locant must be used more than once for the same kind of substituent. The locant immediately precedes the substituent name to which it applies in the name of the compound.

8. Locants occurring together in the name are separated from each other by commas, and all locants are separated from the rest of the name by hyphens.

9. The position of free valence (point of attachment) of a hydrocarbon group may be associated with any locant of the parent chain in the new IUPAC style, but it is always on position 1 in traditional IUPAC style names. For example, $CH_3 — CH_2 — \overset{|}{CH} — CH_3$ may be named by either of the systematic names

butan-2-yl or 1-methylpropyl (the only one used by *Chemical Abstracts* for indexing) but <u>not</u> 2-butyl.

SUBSTITUTIVE NAME PREFIXES AND SUFFIXES FOR SOME IMPORTANT FUNCTIONAL GROUPS

(Arranged in Descending Order of Preference for Citation as Suffixes)

Class	Formula of Group[a]	Prefix[b]	Suffix[c]
Radicals			-yl
Anions			-ide
Cations			-ylium
Carboxylic acids	—COOH	carboxy	-carboxylic acid
	—(C)OOH	—	-oic acid
Sulfonic acids	—SO_3H	sulfo	-sulfonic acid
Esters	—COOR	R^d-oxycarbonyl	R^d...-carboxylate
	—(C)OOR	—	R^d...-oate
Acid halides	—CO—Hal	halocarbonyl	-carbonyl halide
	—(C)O—Hal	—	-oyl (or -yl) halide
Amides	—CO—NH_2	aminocarbonyl or carbamoyl	-carboxamide
	—(C)O—NH_2	—	-amide
Nitriles	—C≡N	cyano	-carbonitrile
	—(C)≡N	—	-nitrile
Aldehydes	—CHO	formyl	-carbaldehyde[e]
	—(C)HO	oxo	-al
Ketones	>(C)=O	oxo	-one
Alcohols	—OH	hydroxy	-ol
Phenols	—OH	hydroxy	-ol
Thiols	—SH	sulfanyl	-thiol
Amines	—NH_2	amino or azanyl	-amine
Ethers	—OR	R^d-oxy	—
Sulfides	—SR	R^d-sulfamyl	—

[a] C in parentheses is included in the stem of the parent chain and not in the prefix or suffix.

[b] Functional group is treated as a substituent.

[c] Functional group is part of parent compound; suffix is added to name of corresponding hydrocarbon.

[d] R is alkyl, aryl, etc.; when R is part of a prefix, the name of the R group is written as part of the prefix name without a hyphen (except with locants).

[e] *Chemical Abstracts* uses -carboxaldehyde in indexes.

NAMES OF SOME IMPORTANT PARENT COMPOUNDS NOT SPECIFICALLY INCLUDED IN THIS PROGRAM

The rules of nomenclature included in this program are readily applied to substituted compounds based on the following parent compounds. Fixed numbering in these parent compounds is indicated by the locants with the formulas.

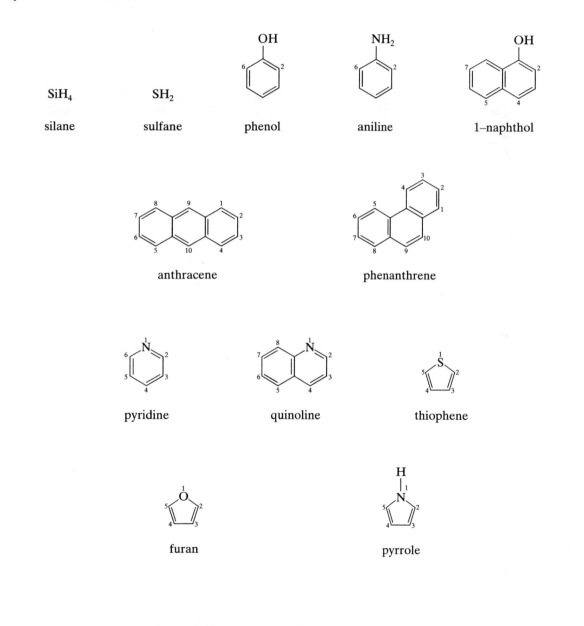

REPLACEMENT NAME PREFIXES FOR HETERO ATOMS[a]
(Listed in Descending Order of Precedence)

Element	Prefix
oxygen	oxa
sulfur	thia
nitrogen	aza
phosphorus	phospha
silicon	sila
boron	bora

[a] For a more complete list, see either the IUPAC or Fletcher
entry in the Selected References.

Answer Sheets

IMPORTANT NOTE

Remove the answer sheet for the chapter being studied by tearing it from the book along the perforated line. Cover the answers with an index card or a sheet of paper, and <u>after</u> you have written your answer in the appropriate blank expose the answers one at a time to check your response.

CHAPTER 1—ALKANES

1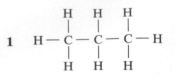

2. four
    one

3. $CH_3—CH_2—CH_3$

4. four
    four

5. meth
    ane

6. ethane
    propane

7. $CH_3 — CH_2 — CH_2 — CH_3$

    $(or\ CH_3 — \overset{\displaystyle CH_3}{\underset{\displaystyle |}{CH}} — CH_3)$

8. alkane

9. seven

10. $\cancel{CH_3 — CH_2 — CH} — \overset{\displaystyle |}{\underset{\displaystyle CH_3}{}}CH — CH_2 — CH_3$
    with $\underset{\displaystyle CH_3}{\cancel{CH} — CH_2 — CH_3}$

11. hept
    heptane

12. three

13. meth
    yl
    two
    ethyl
    alkyl

14. methyl
    methyl
    ethyl

15. dimethyl

16. ethyldimethylheptane

17. eleven
    seven
    two, one, one (or one, one, two)
    eleven

18. 3, 4, 5

19. 4-ethyl-3,5-dimethylheptane

20. the longest continuous chain of carbon atoms,
    or the parent chain, or the chain of carbon
    atoms serving as a basis of the name
    substituents
    three
    positions of substituents along the parent
    (longest continuous) chain
    commas
    hyphens

21. longest continuous
    seven
    hept
    heptane
    three
    methyl, ethyl, ethyl

22. 2, 4, 5
    3, 4, 6
    2, 4, 5
    2; 4 and 5
    commas; hyphens

23. 4,5-diethyl-2-methylheptane

24. five
    pentane
    2,2,4-trimethylpentane

25. 5,5-diethyldecane

26. $CH_3 — CH_2 — \overset{\displaystyle CH_3}{\underset{\displaystyle \underset{\textstyle CH_3}{|}}{C}} — \overset{}{\underset{\displaystyle \underset{\textstyle CH_2}{|}\ CH_3}{CH}} — CH_2 — CH_3$

27. $CH_3 — \overset{\displaystyle CH_3}{\underset{\displaystyle |}{CH}} — CH_2 — \overset{\displaystyle CH_3}{\underset{\displaystyle |}{CH}}$
    $CH_3 — CH_2 — CH_2 — \underset{\displaystyle \underset{\textstyle CH_3}{|}}{CH}$

28. C_6H_{14}
    C_6H_{14}
    isomers

29. $CH_3 — CH_2 — CH_2 — CH_2 — CH_3$

    $CH_3 — \overset{\displaystyle CH_3}{\underset{\displaystyle |}{CH}} — CH_2 — CH_3$

    $CH_3 — \overset{\displaystyle CH_3}{\underset{\displaystyle \underset{\textstyle CH_3}{|}}{C}} — CH_3$

30. four
one

31. pentane
2-methylbutane
2,2-dimethylpropane (locants unnecessary in the last two names because substituent methyls cannot be on any other carbon, but *Chemical Abstracts* uses them)

32. $CH_3 - CH_2 - CH_2 - CH_3$

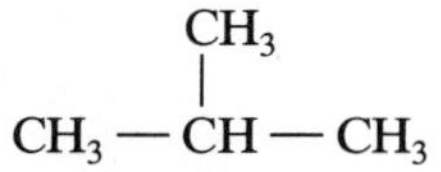

$$CH_3 - \underset{\underset{\displaystyle CH_3}{|}}{CH} - CH_3$$

2-methylpropane (locant unnecessary, but used by *Chemical Abstracts*; see last response in item 31)

33. pent
5
iso
isopentane

34. hexane
isohexane
2-methylpentane

35. 3-methylpentane

36. $CH_3 - CH_2 - CH_2 - CH_2 - CH_3$

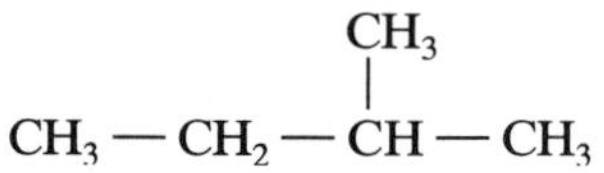

$$CH_3 - CH_2 - \underset{\underset{\displaystyle CH_3}{|}}{CH} - CH_3$$

37. $CH_3 - \underset{\underset{\displaystyle CH_3}{|}}{\overset{\overset{\displaystyle CH_3}{|}}{C}} - CH_3$

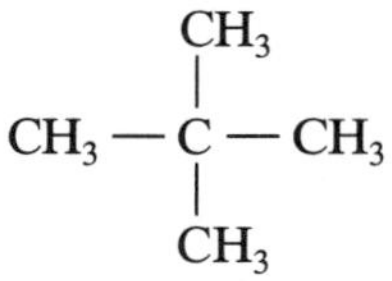

38. neopentane

39. 2,2-dimethylbutane

40. 3,3-dimethylhexane

41. C_4H_8

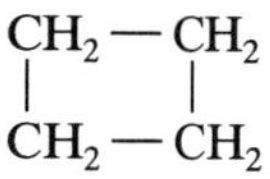

$$\begin{array}{ccc} CH_2 & - & CH_2 \\ | & & | \\ CH_2 & - & CH_2 \end{array}$$

cyclobutane

42. cyclohexane

43. ethylcyclopentane

44. ethyl
1
3
1-ethyl-3-methylcyclohexane

45. 1,3-dimethylcyclobutane

46. ethyl
1,2,4,7
1,2,3,5
2-ethyl-1,3,5,-trimethylcycloheptane

47.

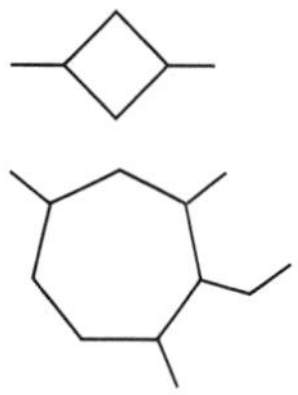

48. 2-chloro-1,3-diethyl-4-methylcyclopentane

CHAPTER 2—NOMENCLATURE OF ALKYL GROUPS

1. methyl
ethyl

2. 5
3
1,1,4
1,1-diethyl-4-methylpentyl

3. 1-ethylbutyl
1-ethylbutylcycloheptane

4. 1-methylbutyl
2-methylbutyl
2,2-dimethylpropyl

5. 1,1,3-trimethylpentyl
1-methylpropyl
1-methylpropyl
1
4
1-(1-methylpropyl)-4-(1,1,3-trimethylpentyl)cyclooctane

6. but
butyl
primary

7. secondary

8. primary
tertiary

9. primary

10. $CH_3 - CH_2 - CH - CH_3$
                        |

11.
$$CH_3 - \underset{\underset{CH_3}{|}}{\overset{\overset{CH_3}{|}}{C}} -$$

tert-butyl

12. tert-pentyl
1,1-dimethylpropyl

13.
$$CH_3 - \underset{}{\overset{\overset{CH_3}{|}}{CH}} - CH_3$$
$$CH_3 - \underset{}{\overset{\overset{CH_3}{|}}{CH}} - CH_2 -$$
primary

14.
$$CH_3 - \underset{}{\overset{\overset{CH_3}{|}}{CH}} - CH_2 - CH_3$$
$$CH_3 - \underset{}{\overset{\overset{CH_3}{|}}{CH}} - CH_2 - CH_2 -$$
primary

15. isohexyl

16.
$$CH_3 - \underset{\underset{CH_3}{|}}{\overset{\overset{CH_3}{|}}{C}} - CH_3$$
$$CH_3 - \underset{\underset{CH_3}{|}}{\overset{\overset{CH_3}{|}}{C}} - CH_2 -$$

17. branched

18. $CH_3 - CH_2 - \overset{|}{CH} - CH_3$

19. $CH_3 - \underset{\underset{CH_3}{|}}{CH} - CH_2 -$

20. primary
secondary

21. 1-methylpropyl
2-methylpropyl
1,1-dimethyllethyl

22. secondary
isopropyl

23. isopropyl
isopropylcyclobutane

24. 10
methyl and sec-butyl
2 and 5
sec-butyl
5-sec-butyl-2-methyldecane

25. 10
isobutyl
5
decane
5-isobutyldecane

26. nonane
ethyl, ethyl, and 1,2-dimethylpropyl
e
d
5
5-(1,2-dimethylpropyl)-3,7-diethylnonane

27. 10
5
2,4,5,7,8
methyl, propyl, and butyl
5-butyl-2,8-dimethyl-4,7-dipropyldecane

28. isopropyl
isopropyl chloride
2-chloropropane

29. $CH_3-CH_2-CH_2-CH_2-Cl$

$CH_3-CH_2-\underset{\underset{Cl}{|}}{CH}-CH_3$

$CH_3-\underset{\underset{CH_3}{|}}{\overset{\overset{CH_3}{|}}{C}}-Cl$

30. $CH_3-\underset{\underset{CH_3}{|}}{\overset{\overset{CH_3}{|}}{C}}-CH_2-Cl$

1-chloro-2,2-dimethylpropane
primary

31. 1-chloro-3-methylbutane
isopentyl chlordie
primary

32. pentane
5
$C-C-C-C-C$
3
2, 2, and 4

$\overset{1}{C}-\overset{2}{C}-\overset{3}{C}-\overset{4}{C}-\overset{5}{C}$

$C-\underset{\underset{CH_3}{|}}{\overset{\overset{CH_3}{|}}{C}}-C-\underset{\underset{CH_3}{|}}{C}-C$

$CH_3-\underset{\underset{CH_3}{|}}{\overset{\overset{CH_3}{|}}{C}}-CH_2-\underset{\underset{CH_3}{|}}{CH}-CH_3$

33. $CH_3-CH-\underset{\underset{\underset{\underset{CH_3}{|}}{CH-CH_3}}{|}}{\overset{\overset{Cl}{|}}{C}}-\overset{\overset{CH_3}{|}}{CH}-CH_2-CH_2-CH_3$

34. $CH_3-\underset{\underset{Cl}{|}}{CH}-CH_2-CH_2-CH_3$

$CH_3-CH_2-\underset{\underset{Cl}{|}}{CH}-CH_2-CH_3$

$CH_3-\underset{\underset{Cl}{|}}{CH}-\underset{\underset{CH_3}{|}}{CH}-CH_3$

35. $CH_3-\underset{\underset{CH_3}{|}}{\overset{\overset{Cl}{|}}{C}}-CH_2-CH_2-CH_3$

$CH_3-CH_2-\underset{\underset{CH_3}{|}}{\overset{\overset{Cl}{|}}{C}}-CH_2-CH_3$

$CH_3-\underset{\underset{CH_3}{|}}{\overset{\overset{Cl}{|}}{C}}-\overset{\overset{CH_3}{|}}{CH}-CH_3$

36. 2-chloro-3-methylbutane

37. $CH_3-\underset{\underset{CH_3}{|}}{\overset{\overset{Cl}{|}}{C}}-\overset{\overset{CH_3}{|}}{CH}-CH_3$

CHAPTER 3—NOMENCLATURE OF ALKENES

1. seven

2. heptene

3. <u>sec</u>-butyl (<u>or</u> 1-methylpropyl)

4. 2
   3

5. 2-heptene

6. 3-<u>sec</u>-butyl-2-heptene [<u>or</u> 3-(1-methylpropyl)-2-heptene]

7. hexene

8. 1
   3, 5, 5

9. that numbering would incorrectly assign a higher number (5) to the alkene linkage (C = C), which must have the smaller possible number

10. 3,5,5-trimethyl-1-hexene

11. $CH_3 - CH_2 - CH = CH_2$,
    $CH_3 - CH = CH - CH_3$, and
    $CH_3 - C = CH_2$
    $\qquad\qquad |$
    $\qquad\quad CH_3$

12. 1-butene, 2-butene, and 2-methylpropene (In the last name the number is unnecessary because the substituent methyl cannot be on any other carbon if the basis of the name is to be propene).

13. cyclooctene

14.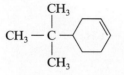

15.

$$CH_3 - \underset{\underset{CH_3}{|}}{\overset{\overset{CH_3}{|}}{C}} - \text{(cyclohexene ring)}$$

16.

(cyclopentene ring) $- CH - CH_2 - CH_3$
$\qquad\qquad\qquad\quad |$
$\qquad\qquad\qquad CH_3$

17. 3
    3-chloropropene (<u>or</u> 3-chloro-1-propene)

18. 1 and 2
    1-chloro-2-isobutylcyclobutene

19. 1 and 5
    2 and 3

1 and 5
1-chloro-5-methylcyclopentene

20. 3,4,4,5,6 and 3,4,5,5,6
    3,4,4,5,6
    3,5-dichloro-4,4,6-trimethylcyclohexene

21. $CH_3 - CH_2 - CH = CH - CH_2 - CH_3$

22. pent-2-ene
    4-methylpent-2-ene

23. 4-chlorobut-1-ene
    5-chloro-3-methylpent-2-ene

24.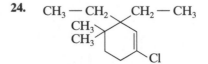

25. 1-propenyl

26. $CH_2 = CH -$

27. $CH_2 = CH - Cl$

28. $CH_2 = CH - CH_2 - Cl$

29. primary

30. vinyl
    4
    4-vinylcyclohexene

31. allyl bromide

32. 2-methyl-2-propenyl and 2-methylallyl

33. 1-methyl-1-propenyl

34.

    tertiary

35. 2,4-dimethylpent-3-en-1-yl

36. 4,4-dimethylcyclohex-2-en-1-yl
    4-(4,4-dimethylcyclohex-2-en-yl)dec-3-ene
    the decene parent chain
    the cyclohexene ring

37. $CH_3 - CH_2 - CH = CH_2$,
    $CH_3 - CH = CH - CH_3$, and
    $CH_3 - C = CH_2$
    $\qquad\qquad |$
    $\qquad\quad CH_3$

38. one
    2-butene <u>or</u> but-2-ene

39.

$$\underset{H}{\overset{CH_3}{>}} C = C \underset{H}{\overset{CH_3}{<}} \quad \underline{and} \quad \underset{H}{\overset{CH_3}{>}} C = C \underset{CH_3}{\overset{H}{<}}$$

40.

$$CH_3 \diagdown C = C \diagup CH_3 \\ H \diagup \quad \diagdown H$$

41.

$$CH_3 \diagdown C = C \diagup H \\ H \diagup \quad \diagdown CH_3$$

42.

$$CH_3 \diagdown C = C \diagup H \\ H \diagup \quad \diagdown CH_2 - CH_3$$

43. the parent chain extends from opposite sides of the alkene linkage

44. 5-methyl-_cis_-hex-2-ene _or_ 5-methyl-_cis_-2-hexene

45.

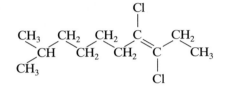

46.

47. trans
10
trans-cyclodecene

48. Cl, H
higher atomic number gives higher priority (rule 1)
higher atomic numbers of second atoms out give
higher priority (rule 3)
$CH_3 - CH_2, CH_3 -$
opposite
E
1-butene
(_E_)-1-chloro-2-methyl-1-butene

49. C, H, H
C, C, H
H, H, H
lower
higher
E
(_E_)-4-ethyl-3,5-dimethyl-3-heptene
trans

50. E
cis

51.

$$Cl \diagdown C = C \diagup \triangle \\ CH_3 - CH_2 \diagup \quad \diagdown H$$

52. E

53. penta-1,4-diene
$CH_2 = CH - CH_2 - CH = CH_2$
penta-1,3-diene
$CH_2 = CH - CH = CH - CH_3$

54. 1,3-butadiene
buta-1,3-diene

55.

$$CH_2 = C - CH = CH_2 \\ | \\ CH_3$$

56.

$$CH_3 \diagdown CH \diagdown C = C \diagup CH_2 - CH_2 \diagdown C = C \diagup CH_3 \\ CH_3 \diagup \quad | \quad CH_2 - CH_2 \diagup \quad \diagdown H \\ \quad \quad Cl \quad | \\ \quad \quad \quad CH_3$$

(H above on first C)
isolated

57. $CH_2 = C = CH_2$

58. 1,2-butadiene

59.

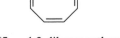

(cyclohexadiene ring structure)

60. conjugated

61. isolated

62. $CH_2 = C = CH_2$
cumulated

63.

$$CH_3 - CH - CH = CH - C = CH \\ | \quad \quad \quad \quad | \\ Cl \quad \quad \quad CH_3 - CH_2$$

(with $CH_3 - CH_2 - CH_2 - CH_2$ above)
conjugated

$$CH_3 - CH - Cl \\ \quad \quad C = C \\ H \diagup \quad \diagdown H$$

(with CH₃—CH₂ above and CH₂—CH₂—CH₂—CH₃ structure)

64.

(cyclooctatetraene ring structure)

65. 1,2-dibromoethane

CHAPTER 4—ALKYNES

1. propyne

2. 7
   hept
   yne
   heptyne
   3
   3-heptyne
   methyl, ethyl
   2, 5
   5-ethyl-2-2methyl-3-heptyne

3. but-2-yne

4. $CH_3 — CH_2 — C \equiv C — CH_2 — CH_3$

5. 8
   oct
   $C \equiv C$
   $C — C \equiv C — C — C — C — C — C$
   Cl 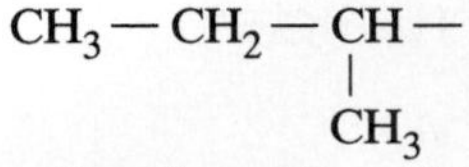 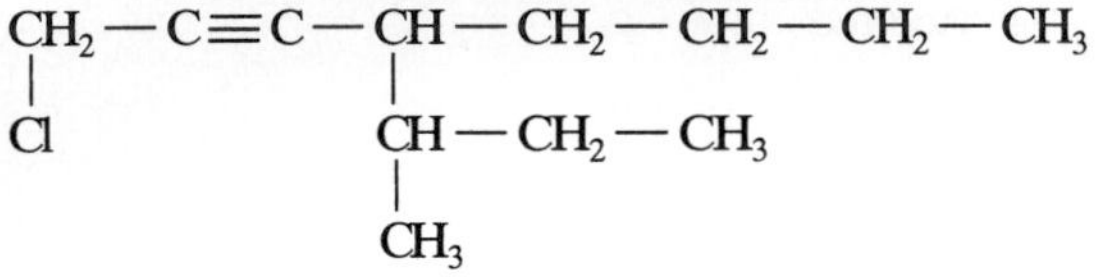
   $CH_3 — CH_2 — CH —$
   |
   CH_3
   $CH_2 — C \equiv C — CH — CH_2 — CH_2 — CH_2 — CH_3$
   | |
   Cl $CH — CH_2 — CH_3$
                              |
                            CH_3
   (1-methylpropyl)
   1-chloro-4-(1-methylpropyl)-2-octyne

6. 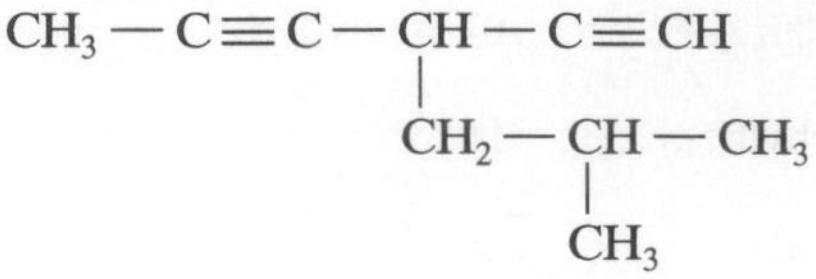 $CH_3 — C \equiv C — CH — C \equiv CH$
   |
   $CH_2 — CH — CH_3$
   |
   CH_3

   3-(2-methylpropyl)hexa-1,4-diyne

7. 1,6-cyclodecadiyne
   3,4,9
   3,4,9-trimethyl-1,6-cyclodecadiyne

8. butenyne (locants are unnecessary, because only one butenyne is possible, but *Chemical Abstracts* uses them: 1-buten-3-yne)

9. 4-chloro-6-isopropyl-6-nonen-2-yne
   6-chloro-4-isopropyl-2-nonen-7-yne

10. 4-ethyl-4-methylhex-1-en-5-yne

11. ethynyl

12.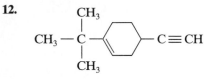

13. $HC \equiv C — CH_2 —$
    $CH_3 — C \equiv C —$

14. $— CH_2 — C \equiv C — CH = CH_2,$
    $CH_3 — C \equiv C — C = CH_2,$ and
    |

    $CH_3 — C \equiv C — CH = CH —$

    4-penten-2-ynyl, (1-propynyl)ethenyl, and 1-penten-3-ynyl

CHAPTER 5—AROMATIC HYDROCARBONS

1. ethylbenzene

2. $H-C \equiv C-$

   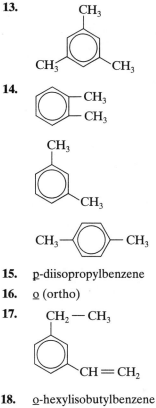

   $-C \equiv CH$

3. CH—CH$_2$—CH$_3$ with CH$_3$ below

4. neopentyl
   neopentylbenzene

5. triphenylmethane

6. $-C \equiv C-H$

   phenylethyne

7. 2-nonene
   4
   6
   4-methyl-6-phenyl-2-nonene or 4-methyl-6-phenyl-non-2-ene

8. CH$_2$—CH$_2$—CH—CH$_2$ ... CH$_3$
   CH$_2$ ... CH$_2$ ... CH$_2$
   CH$_3$ ... CH$_2$—CH$_2$—CH$_2$

   or CH$_3$—(CH$_2$)$_3$—CH—(CH$_2$)$_6$—CH$_3$ with ϕ below

9. $-CH_2-CH=CH_2$
   or C$_6$H$_5$—CH$_2$—CH$=$CH$_2$
   or ϕ—CH$_2$—CH$=$CH$_2$
   3-phenylpropene
   C$_6$H$_5$—CH$=$CH—CH$_3$
   1-phenylpropene

10. 1,3

11. CH$_2$—CH$_3$
    CH$_2$—CH$_3$

12. 1,2,4
    2-isopropyl-1,4-dimethylbenzene

13.

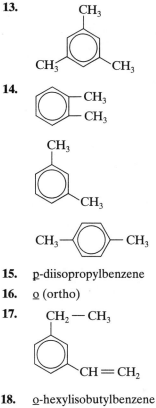

14.

15. p-diisopropylbenzene

16. o (ortho)

17. CH$_2$—CH$_3$

    CH$=$CH$_2$

18. o-hexylisobutylbenzene

19. CH$_2$—CH$_3$

    CH$_3$

20. CH$_3$
    CH$_3$—C—CH$_3$
    CH$_3$

    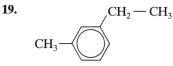

21. 2-allyltoluene or o-allyltoluene or
    2-(2-propenyl)toluene or o-(2-propenyl)toluene

22. 1-sec-butyl-4-methylbenzene or
    1-methyl-4-(1-methylpropyl)benzene
    4-sec-butyltoluene or p-sec-butyltoluene

23. CH$_2$=CH— or CH$_2$=CH—ϕ

24. CH$_2$=CH— CH$_2$—CH—CH$_3$ with CH$_3$ below

25. (1-methylethenyl)benzene

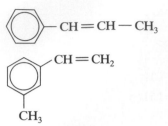

26. 2-phenyl-1-pentene

27.

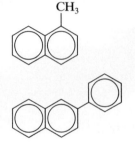

28. 2-tert-butyl-2',6-dimethylbiphenyl

29.

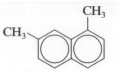

30. 2-methyl-4-(2-naphthyl)-1-butene
2-(3-methyl-3-butenyl)naphthalene

31. 1,4
1,4-dihydronaphthalene
1,4,5,8
1,4,5,8-tetrahydronaphthalene
1,2,3,4-tetrahydronaphthalene

32.

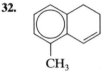

33.

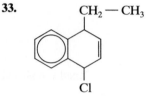

34. 4,5-dimethyl-3-(pentan-2-yl)-1,2-dihydronaph-thalene

CHAPTER 6—ALCOHOLS

1. isopropyl alcohol

2. $CH_3 - CH - CH_2 - CH_3$
$|$
OH

3. alcohol is a class name rather than the name of a specific compound

4. isopentyl
isopentyl alcohol

5. allyl
allyl alcohol

6.
CH_3
$|$
$CH_3 - C - OH$
$|$
CH_3

7. $CH_3 - CH_2 - OH$
primary

8. $CH_3 - CH - CH_2 - OH$
$|$
CH_3
primary

9. primary

10. the OH group is attached to a primary carbon, that is, to a carbon that is attached to only one other carbon

11. secondary

12. $CH_3 - CH - CH_2 - CH_3$
$|$
OH
sec-butyl alcohol

13.
OH
$|$
$CH_3 - CH - CH_2 - CH_2 - CH_3,$

OH
$|$
$CH_3 - CH_2 - CH - CH_2 - CH_3,$

OH
$|$
$CH_3 - CH - CH - CH_3$
$|$
CH_3

14.
CH_3
$|$
$CH_3 - C - OH$
$|$
CH_3
tert-butyl alcohol

15. primary

16. 3-pentanol
pentan-3-ol

17.
OH
$|$
$CH_3 - CH - CH_2 - CH_3$

18. cyclohexanol

19. 3-methyl-2-butanol
3-methylbutan-2-ol

20. 9
nonane

21. nonanol

22. 4
4-nonanol
nonan-4-ol

23. 3
chloro, phenyl, and isopropyl
3, 4, and 6

24. hyphens
commas

25. 3-chloro-6-isopropyl-4-phenyl-4-nonanol

26. tertiary

27. $C_6H_5 - CH_2 - OH$
phenylmethanol

28. $CH_3 - CH - CH_2 - OH$
$|$
OH

29. propane-1,3-diol

30. propane-1,2,3-triol

31. 2 and 4
lower possible
5,6-dimethyl-6-phenylheptane-2,4-diol

32. cyclohexane-1,2,4-triol

33. $CH_2 = CH - CH_2 - OH$
2-propen-1-ol <u>or</u> prop-2-en-1-ol

34.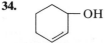

35. cyclooct-4-en-1-ol

36.

$$CH_3-CH_2-\overset{\overset{\displaystyle CH_3}{|}}{\underset{\underset{\displaystyle OH}{|}}{C}}-C\equiv CH$$

37. 7
heptenol
3, 5
5-hepten-3-ol <u>or</u> hept-5-en-3-ol
2
methyl, ethyl
6, 4
4-ethyl-6-methyl-5-hepten-3-ol <u>or</u>
 4-ethyl-6-methylhept-5-en-3-ol

38.

$$CH_3-\overset{\overset{\displaystyle }{|}}{\underset{\underset{\displaystyle OH}{|}}{CH}}-\overset{\overset{\displaystyle }{|}}{\underset{\underset{\displaystyle C_6H_5}{|}}{C}}=CH-\overset{\overset{\displaystyle }{|}}{\underset{\underset{\displaystyle Cl}{|}}{CH}}-CH_3$$

39. 6-methyl-3-heptene-2,5-diol

40. 4-<u>sec</u>-butyl-4-cyclohexene-1,2-diol

41.

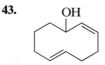

42. 2-methyl-6-phenyl-<u>cis</u>-3-decen-1-ol

43.

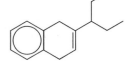

44. 9
2,6,6-trimethyl-1-cyclohexenyl
trans
3,7-dimethyl-9-(2,6,6-trimethyl-1-cyclohexenyl)-
 <u>trans</u>-2-<u>trans</u>-4-<u>trans</u>-6-<u>trans</u>-8-nonatetraen-1-ol

45.

$$HO-CH_2-\overset{\overset{\displaystyle }{|}}{\underset{\underset{\underset{\displaystyle CH_2-OH}{|}}{}}{CH}}-CH_2-OH$$

46. 4-hydroxyphenyl
5-(4-hydroxyphenyl)heptan-2-ol

47.

$$CH_3-CH_2-\overset{\overset{\displaystyle |}{}}{CH}-CH_3$$
1-methylpropyl
<u>sec</u>-butyl

48. 1-ethyl-2,3-dimethylbutyl
4,5-dimethylhexan-3-yl

49. 4-<u>tert</u>-butylcyclohexan-1-yl
4-hydroxy-4-phenylbutan-3-yl

50. 1-methyl-2-propenyl (<u>or</u> 1-methylallyl)
4
but-3-en-2-yl

51.

52.

CHAPTER 7—ETHERS

1. ethoxyethane

2. 7
   heptane
   2, 3, 5
   ethoxy
   5-ethoxy-2,3-dimethylheptane

3.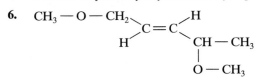

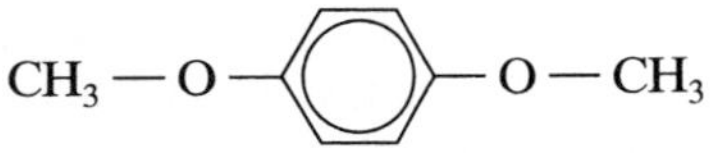

   $CH_3 - CH_2 - CH_2 - CH(O - CH_2 - CH_3)_2$

4. phenoxybenzene

5. heptane
   4
   2, 4, 6
   4-chloro-6-cyclohexyloxy-2,2-dimethylheptane

6.
   $CH_3 - O - CH_2$
   $\underset{H}{} C = C \underset{CH - CH_3}{\overset{H}{}}$
   $O - CH_3$

7. cyclohexanol
   ethoxy, phenyl
   1, 5
   5-ethoxy-2-phenylcyclohexanol

8. 10
   decyne
   2
   isopropoxy
   5
   5-isopropoxy-2-decyne or 5-isopropoxydec-2-yne

9.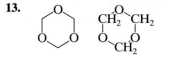
   $C_6H_5 - O - C_6H_5$ or

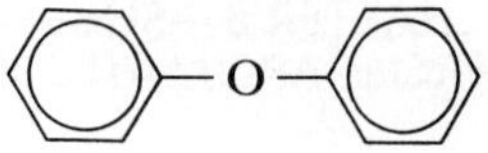

   $CH_3 - \underset{CH_3}{\overset{|}{CH}} - O - C_6H_5$

10. allyl
    diallyl ether

11. 2-sec-butoxybutane or 2-(butan-2-yloxy)butane

12. tert-pentyl
    cyclopropyl
    cyclopropyl tert-pentyl ether
    2-cyclopropoxy-2-methylbutane

13.

14. 10
    3
    3,6,9-trimethyl-2,5,8-trioxadecane

15. 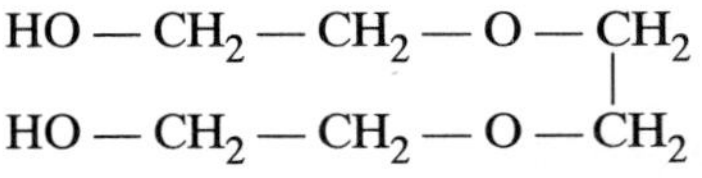 $HO - CH_2 - CH_2 - O - CH_2$
    $HO - CH_2 - CH_2 - O - CH_2$

16. 3,5,8-trioxa-1-nonene or 3,5,8-trioxanon-1-ene

17. 8
    4
    sila
    2,2,7,7-tetramethyl-4-phenyl-3,6-dioxa-2,7-disi-
    laoctane

18. 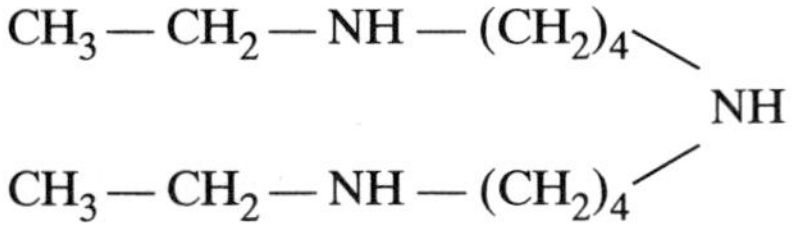 $CH_3 - CH_2 - NH - (CH_2)_4$
    $\qquad\qquad\qquad\qquad NH$
    $CH_3 - CH_2 - NH - (CH_2)_4$

CHAPTER 8—SUBSTITUTION PRODUCTS FROM AROMATIC HYDROCARBONS

1. ethylbenzene
bromobenzene
nitrobenzene
nitrosobenzene

2.
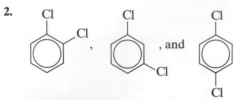

o-dichlorobenzene, m-dichlorobenzene, and p-dichlorobenzene
1,2-dichlorobenzene, 1,3-dichlorobenzene, and 1,4-dichlorobenzene

3.

1-chloro-3-nitrobenzene

4. 1, 2, 3, 5
2-chloro-1,3,5-trinitrobenzene

5.
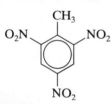

6. the use of benzene as the basis of a name requires the use of the smallest possible numbers,
while the use of toluene as the basis of the name
requires that the methyl group be on position
number 1.
2-methyl-1,3,5-trinitrobenzene

7.
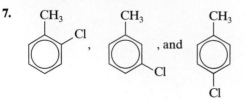

o-chlorotoluene, m-chlorotoluene, and p-chloro-
toluene (or 2-chlorotoluene, 3-chlorotoluene,
and 4-chlorotoluene)

8.

![benzene ring with CH₂—Cl]

9. (chloromethyl)benzene

10.

![benzene ring with CH₂—NO₂]

(nitromethyl)benzene
without them, chloromethyl or nitromethyl
might be mistakenly taken to indicate two substituents on the benzene ring

11. benzyl chloride
benzyl iodide
dibenzyl ether

12.

![benzene ring with CH₂—OH]

primary

13.

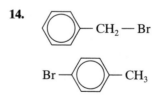

or C_6H_5 — MgBr or Ph — MgBr

![benzene ring with CH₂—MgBr or ∅—CH₂—MgBr]

or C_6H_5 — CH₂ — MgBr or Ph — CH₂ — MgBr

14.

![benzene ring with CH₂—Br]

![Br—benzene ring—CH₃]

15.

![benzene ring with CH₂=CH₂]

16.

![O₂N—benzene ring—CH=CH₂]

1-ethenyl-4-nitrobenzene

17.
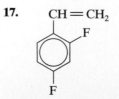

18. benzyl
4-benzyl-3-fluorostyrene

19.

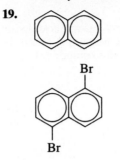

20. 1-methyl-2,4-dinitronaphthalene

21. hydrogens added to the parent, naphthalene
5-ethoxy-1,2,3,4-tetrahydronaphthalene

22.

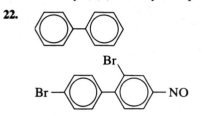

CHAPTER 9—ACIDS

1. $CH_3 - CH_2 - CH_2 - COOH$

$CH_3 - CH = CH - COOH$

2. 6
hex
methyl
3
3-methylhexanoic acid

3. dec-9-enoic acid

4. 7
3 and 5
3-benzyl-5-methyl-3-nitroheptanoic acid

5. $HC \equiv C-$

$CH_3 - CH_2 - O -$

$CH_2 - CH_2 - CH_2 - CH - COOH$

$\overset{|}{O} - CH_2 - CH_3 \qquad \overset{|}{C} \equiv CH$

6.

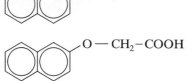

7. 5-phenylpent-4-ynoic acid

8. hexanoic acid
ethoxy and 2,4-dibromophenyl
4-(2,4-dibromophenyl)-3-ethoxyhexanoic acid

9. alkenoic
alkynoic

10. the extension of the parent chain from the alkene linkage

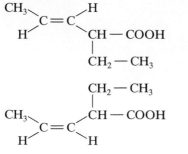

11. 6
$HOOC - CH_2 - CH_2 - CH_2 - CH_2 - COOH$

12. 2,3-dibromo-2,3-diphenylbutanedioic acid

13. <u>E</u>
but-2-enedioic acid
(<u>E</u>)-2-chloro-3-methylbut-2-enedoic acid (locant for C = C not required, because it cannot be anything other than 2 in this compound)

14.

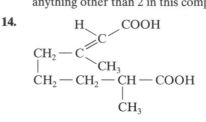

15.

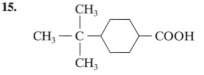

16. cyclooct-1-ene-1-carboxylic acid
cyclooct-4-ene-1-carboxylic acid

17. 9

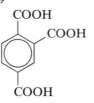

18. naphthalene-2,7-dicarboxylic acid
6-nitrobiphenyl-2,2'-dicarboxylic acid

19.

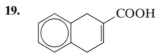

20. 7

$\overset{HOOC}{\underset{HOOC}{\Large >}} CH - CH_2 - CH \overset{COOH}{\underset{COOH}{\Large <}}$

21. pentane-1,2,4,5-tetracarboxylic acid

22.

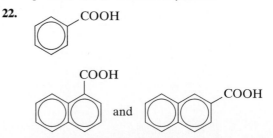

23.

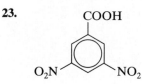

24. 2
6-nitronaphthalene-2-carboxylic acid
6-nitro-2-naphthoic acid

25. benzoic acid
1-propynyl
4-(1-propynyl)benzoic acid

26. prop
propion
but
butyr

27. $CH_3 — COOH$
acetic acid

28. 2

29. chloroacetic acid
trichloroacetic acid

30.

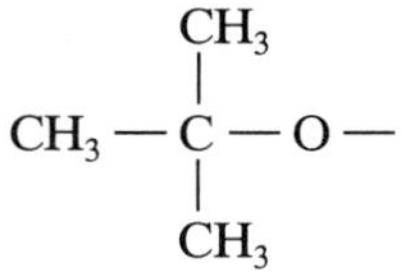

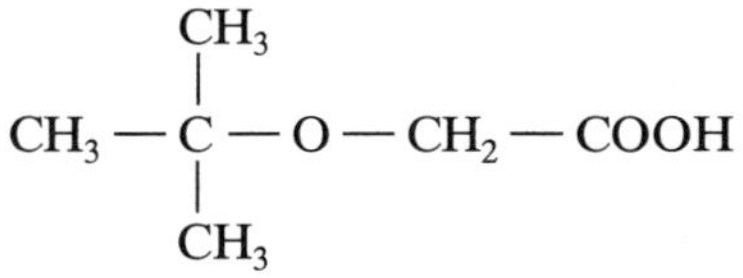

$$CH_3 — \overset{\displaystyle CH_3}{\underset{\displaystyle CH_3}{C}} — O —$$

$$CH_3 — \overset{\displaystyle CH_3}{\underset{\displaystyle CH_3}{C}} — O — CH_2 — COOH$$

31. H — COOH
formic acid

32. peroxyformic acid <u>or</u> peroxymethanoic acid

33.

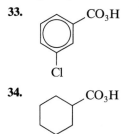

34.

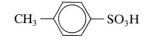

35. trifluoroperoxyacetic acid

36. $C_6H_5 — SO_3H$

$$CH_3 — \langle \bigcirc \rangle — SO_3H$$

37. naphthalene
2
5-nitronaphthalene-2-sulfonic acid

38. trifluoromethanesulfonic acid

39. 3,3-dimethylcyclobutane-1-sulfonic acid

CHAPTER 10—ACID DERIVATIVES

1. acetic acid
   acetic anhydride

2. $C_6H_5 — COOH$
   $C_6H_5 — CO — O — CO — C_6H_5$

3. $CF_3 — CO — O — CO — CF_3$

4. cyclohexanecarboxylic acid
   cyclohexanecarboxylic anhydride

5. trifluoromethanesulfonic acid
   trifluoromethanesulfonic anhydride

6.

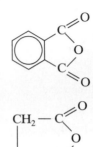

7.

8. naphthalene-1,8-dicarboxylic acid
   naphthalene-1,8-dicarboxylic anhydride

9. benzoic acid and acetic acid
   acetic benzoic anhydride (alphabetical)

10. butyric p-toluenesulfonic anhydride or butanoic
    4-methylbenzene-1-sulfonic anhydride

11. benzoic acid
    benzoyl chloride

12. $CH_3 — COOH$
    $CH_3 — CO — Cl$

13. 2-bromobutyric acid (or 2-bromobutanoic acid)
    2-bromobutyryl bromide (or 2-bromobutanoyl bromide)

14. $O_2N —$ ⟨benzene ring⟩ $— SO_2 — Cl$

15. cyclopropanecarbonyl chloride

16. 2-naphthoyl chloride
    naphthalene-2-carbonyl chloride

17. formic acid
    methanoate
    formate

18. propanoic acid
    propionate
    propanoate

19. butyrate

20. $CH_3 — CH_2 — CH_2 — CO — OCH_3$

21. sec-butyl
    cyclopentanecarboxylic acid
    cyclopentanecarboxylate
    sec-butyl cyclopentanecarboxylate

22. $CF_3 — COOH$
    $CH_3 — CH — CH_2 —$
    $\qquad\quad |$
    $\qquad\quad CH_3$
    $CF_3 — COO — CH_2 — CH — CH_3$
    $\qquad\qquad\qquad\qquad\quad |$
    $\qquad\qquad\qquad\qquad\quad CH_3$

23. $CH_3 — CH — COOH$
    $\qquad\quad |$
    $\qquad\quad CH_3$
    $CH_3 — CH — COO — CH_2 — CH — CH_3$
    $\qquad\quad |\qquad\qquad\qquad\qquad |$
    $\qquad\quad CH_3\qquad\qquad\qquad CH_3$

24. allyl formate

25.
$$C_6H_5 — O — \overset{\overset{\textstyle O}{\|}}{C} — CH_2 — CH_3$$
$$C_6H_5 — CH_2 — O — \overset{\overset{\textstyle O}{\|}}{C} — CH_3$$

26. right
    5
    2,4
    methoxy, methyl
    2-methoxy-4-methylpentyl or 2-methoxy-4-methylpentan-1-yl
    7
    5
    3
    5-chloro-3-methylhept-5-enoic acid
    5-chloro-3-methylhept-5-enoate
    2-methoxy-4-methylpentyl 5-chloro-3-methylhept-5-enoate

27. 3,5-dinitrobenzoate
    2-chloro-4-methylcyclohexyl
    2-chloro-4-methylcyclohexyl 3,5-dinitrobenzoate

28.

$$CH_3-CH_2 \quad \overset{\displaystyle}{\underset{H}{C}}=\overset{\displaystyle H}{\underset{CH_2-O-SO_2}{C}}\!\!-\!\!\langle\text{ring}\rangle\!-\!CH_3$$

29. cyclooct-3-en-1-yl formate

30. $(CH_3-CH_2-CH_2-CH_2-O)_3B$

31. benzenesulfonic acid
benzenesulfonate
4-<u>tert</u>-butylcyclohexyl benzenesulfonate

32. $CH_2=CH-O-SO_2-CF_3$

33. sulfuric
$CH_3-(CH_2)_{10}-CH_2-O-SO_3^-Na^+$

34.

$$CH_3-(CH_2)_3-\overset{\displaystyle}{\underset{H}{C}}=\overset{\displaystyle}{\underset{H}{C}}-(CH_2)_5-CH_2-O-\overset{\displaystyle O}{\overset{\|}{C}}-CH_3$$

35. 3,6-dioxaheptyl <u>or</u> 3,6-dioxaheptan-1-yl
3,6-dioxaheptyl acetate <u>or</u> 3,6-dioxaheptan-1-yl acetate

36. ethoxycarbonyl
4-ethoxycarbonylcyclohexanecarboxylic acid

37.

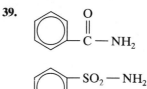

38. formic acid
formamide

39.

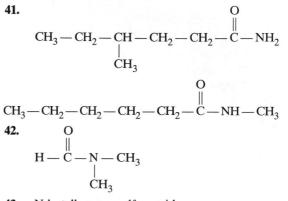

40. 4-methylcyclohexanecarboxylic acid
4-methylcyclohexanecarboxamide

41.

$$CH_3-CH_2-\overset{\displaystyle}{\underset{\underset{CH_3}{|}}{CH}}-CH_2-CH_2-\overset{\displaystyle O}{\overset{\|}{C}}-NH_2$$

$$CH_3-CH_2-CH_2-CH_2-CH_2-\overset{\displaystyle O}{\overset{\|}{C}}-NH-CH_3$$

42.

$$H-\overset{\displaystyle O}{\overset{\|}{C}}-\overset{\displaystyle}{\underset{\underset{CH_3}{|}}{N}}-CH_3$$

43. <u>N</u>-butylbenzenesulfonamide

44. benzamide
<u>N</u> and 3
<u>N</u>,3-dibromobenzamide

CHAPTER 11—ALDEHYDES AND KETONES

1. a ketone

2. an aldehyde

3. an aldehyde
a ketone

4. pentane
pentanal

5.

6. $CH_3 — CH_2 — CHO$

7. $CH_3 — CH_2 — CH_2 — CO — CH_3$
$CH_3 — CH_2 — CO — CH_2 — CH_3$
2-pentanone or pentan-2-one
3-pentanone or pentan-3-one

8. $CH_3 — CH_2 — CH — CO — CH_3$
                                    |
                                   Cl

$CH_3 — CH_2 — CH — CH_2 — CHO$
                            |
                           Cl

9. 3,3-dimethylbutan-2-one (Locant 2 not necessary but generally used.)

10. 5-phenylhex-2-ene
5-phenylhex-2-enal

11. 2-phenylhept-2-ene
1
5
6
6-phenylhept-5-enal

12. heptan-2-one
3,7-dimethyloct-6-enal

13. cyclohex-2-enone
isopropenyl (or 1-methylethenyl)
methyl
2-methyl-5-(1-methylethenyl)cyclohex-2-enone
or 5-isopropenyl-2-methylcyclohex-2-enone

14. $CH_2 = C = O$

15. 9
2,6,6-trimethylcyclohex-1-en-1-yl
3, 7
2-_trans_,4-_trans_,6-_trans_,8-_trans_-nona-2,4,6,8-tetraenal [(_E_) may replace _trans_ each time here.]
3,7-dimethyl-9-(2,6,6-trimethylcyclohex-1-en-l-yl)-2-_trans_,4-_trans_,6-_trans_,8-_trans_-nona-2,4,6,8-tetraenal [(_E_) may replace _trans_ each time here.]

16. 1
cyclohexyl
1-cyclohexyl-2-pentanone

17.
$C_6H_5 — CO — CH_2 — CH_2 — CH — CH_2 — CH_3$
                                              |
                                            C_6H_5

18. 1-cyclobutyl-2,3-dimethylbutan-1-one

19. $OHC — CH_2 — CH_2 — CH_2 — CH_2 — CHO$
$CH_3 — CO — CH_2 — CO — CH_2 — CH_3$

20. 4-phenylheptanedial

21. cyclopentanecarbaldehyde
3-methylcyclobutanecarbaldehyde

22.

23. 4-nitrobenzenecarbaldehyde

24. $CH_3 — COOH$
$CH_3 — CHO$

25. isobutyric acid
isobutyraldehyde

26. $C_6H_5 — CHO$
benzoic acid
benzaldehyde

27. trichloroacetaldehyde

28. $CH_3 — CH_2 — CO — CH_3$

29. dicyclopropyl ketone
diphenyl ketone

30. ethyl 2-naphthyl ketone
1-(2-naphthyl)-1-propanone or 1-(naphthalen-2-yl)propan-1-one

31. $C_6H_5 — CH_2 — CO — CH — CH_2 — CH_3$
                                      |
                                    CH_3

3-methyl-1-phenyl-2-pentanone

32. benzophenone

33.

$CD_3 — CO — CD_3$

34.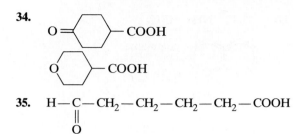

35. $\text{H} - \overset{\displaystyle \text{O}}{\underset{\displaystyle \|}{\text{C}}} - \text{CH}_2 - \text{CH}_2 - \text{CH}_2 - \text{CH}_2 - \text{COOH}$

an aldehyde

36. $\text{CH}_3 - \overset{\displaystyle \text{O}}{\underset{\displaystyle \|}{\text{C}}} - \text{CH}_2 - \text{CH}_2 - \text{CH}_2 - \text{CH}_2 - \text{CH}_2 - \overset{\displaystyle \text{H}}{\underset{\displaystyle \text{H}}{\text{C}}} = \text{C} - \text{COOH}$

37. cyclodecenecarbaldehyde
2
7-oxocyclodec-2-enecarbaldehyde

38. vinyl 3-formylcyclopentanecarboxylate or ethenyl
3-formylcyclopentanecarboxylate

39.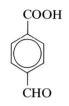

CHAPTER 12—AMINES AND RELATED CATIONS

1. secondary

2. tertiary

3. $R—NH_2$
   $R—NH—R$ <u>or</u> R_2NH
   $R—N—R$ <u>or</u> R_3N
   $\quad\quad |$
   $\quad\quad R$

4. tertiary
   primary

5. secondary
   primary

6. primary
   primary

7. butan-2-amine <u>or</u> 2-butanamine
   butan-2-ylamine <u>or</u> <u>sec</u>-butylamine
   $\quad$<u>or</u> (1-methylpropyl)amine
   butan-2-ylazane <u>or</u> <u>sec</u>-butylazane
   $\quad$<u>or</u> (1-methylpropyl)azane

8. $\quad\quad\quad CH_3$
   $\quad\quad\quad\;|$
   $CH_3—C—NH_2$ <u>or</u> $(CH_3)_3C—NH_2$
   $\quad\quad\quad\;|$
   $\quad\quad\quad CH_3$

9. diphenylamine

10.
   $\quad\quad\quad\quad\quad\quad\quad\quad CH_2—CH_3$
   $\quad\quad\quad\quad\quad\quad\quad\quad\quad\;\;|$
   $CH_3—CH_2—CH_2—CH_2—CH—NH_2$

11. 3-ethyl-3,5-dimethylhexan-1-amine

12. 4-isopropyl-2-phenylcyclohexanamine

13.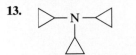

14. 2-methylpropan-1-amine (<u>or</u> isobutylamine)
    <u>N</u>-isopropyl-2-methylpropan-1-amine

15. benzene
    <u>N,N</u>-dimethylbenzenamine

16. $CH_2{=}CH{-}\bigcirc{-}N(CH_3)_2$

    NH_3, amine

17.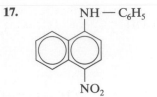

18. 2
    5-methylhexan-2-amine

19. octene
    <u>N,N</u>-dimethyl-6-phenoxyoct-4-en-1-amine

20. 6-methoxycyclooct-3-en-1-amine

21. <u>N</u> and 4
    <u>N</u>
    <u>N</u>-ethyl-<u>N</u>,4-dimethylhex-2-en-2-amine

22. amino
    4-aminohexan-3-ol

23. 4
    dimethylamino
    5
    5-dimethylamino-4-methylpentan-2-one

24.
   $\quad\quad\quad\quad\quad\quad\quad\quad\quad\quad O$
   $\quad\quad\quad\quad\quad\quad\quad\quad\quad\quad\|$
   $CH_3—CH_2—O\quad\;C—O—CH_2—CH_3$
   $\quad\quad\quad\quad\quad\quad\;|$
   $CH_3—(CH_2)_6—CH—CH—N(CH_2—CH_3)_2$

25. butane-1,4-diamine
    pentane-1,5-diamine

26.
   $H_2N—CH_2—CH_2—CH_2—CH_2—CH_2—CH_2—NH_2$

27.
   $\quad\quad\quad\quad\quad CH_3$
   $\quad\quad\quad\quad\quad\;|$
   $CH_3—CH—CH—CH—CH_3$
   $\quad\quad\;\;|\quad\quad\quad\quad\;|$
   $\quad\quad NH_2\quad\quad\;\;NH_2$

28. pentane-1,2,5-triamine

29. $(CH_3)_2N—CH_2—CH_2—N(CH_3)_2$
    $H_2N—CH_2—CH_2—NH_2$

30. <u>N</u>-ethylethanamine <u>or</u> diethylamine
    <u>N</u>-ethylethanammonium chloride
    $\quad$<u>or</u> diethylammonium chloride
    $\quad$<u>or</u> <u>N</u>-ethylethanaminium
    chloride

31. $\bigcirc\bigcirc{-}\overset{+}{N}H_3\;\;\overset{-}{Cl}O_4$

32. 4-tert-butyl-N,N-dipropylcyclohexylammonium or
4-tert-butyl-N,N-dipropyl-cyclohexan-1-aminium

33.

pyridinium

34. $HO - \overset{+}{N}H_3 \ \overset{-}{Cl}$
hydroxylammonium chloride

35. $C_6H_5 - CH_2 - \overset{+}{N}(CH_3)_3 \ \overset{-}{Cl}$

36. N-ethyl-N-octyloctan-1-amine
diethyldioctylammonium iodide

37. $R_4\overset{+}{N} \ \overset{-}{OH}$

38. (2-hydroxyethyl)trimethylammonium hydroxide

39. $CH_3 - CH - \overset{-}{COO}$
$\qquad\quad |$
$\qquad\quad \overset{+}{N}H_3$

$CH_3 - CH_2 - \overset{-}{COO} \ NH_4^+$
are not

40. $CH_3 - CH - CH_2 - CH - \overset{-}{COO}$
$\qquad\quad | \qquad\qquad\quad |$
$\qquad\quad CH_3 \qquad\qquad \overset{+}{N}H_3$

41. 2-ammonio-3-hydroxylbutyrate (or 2-ammonio-3-hydroxybutanoate)

42. trimethylammonioacetate or trimethylammonioethanoate)

CHAPTER 13—BRIDGED RING SYSTEMS

1. 8
    bicyclooctane

2. bicyclopentane

3. 3, 2, 1

4. 2, 1, 0

5. bicyclo[2.1.0]pentane

6. bicyclo[4.2.0]octane
    bicyclo[4.1.1]octane
    bicyclo[3.3.0]octane
    bicyclo[5.1.0]octane
    bicyclo[2.2.2]octane

7. 2
    7

8. 7
    3

9.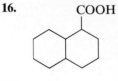

10. CH_3 —⬦— CH_3

11. bicyclo[2.2.1]hept-2-ene

12.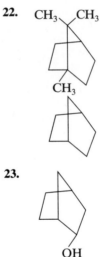
    CH_3 CH_3
    CH_3

13. 3,7,7-trimethylbicyclo[4.1.0]hept-3-ene

14. bicyclo[2.2.1]hexan-5-ol

15. 1,7,7-trimethylbicyclo[2.2.1]heptan-2-one

16. COOH

17.   O

18. 9-oxabicyclo[4.2.1]nonane

19. bicyclo[3.2.2]nonane
    aza
    3-azabicyclo[3.2.2]nonane

20.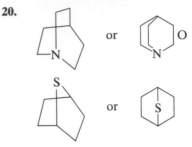

21. 3
    8-azabicyclo[3.2.1]octane-2-carboxylic acid
    methyl 3-benzoyloxy-8-methyl-8-azabicyclo[3.2.1]
    octane-2-carboxylate

22. CH_3 CH_3

    CH_3

23.   OH

24. endo-2-bromobicyclo[2.2.1]heptane

25. exo-bicyclo[2.2.1]heptan-2-amine or exo 2-bicy-
    clo[2.2.1]heptylamine

26.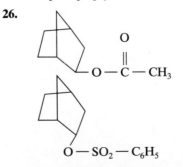
    O
    ‖
    O — C — CH_3

    O — SO_2 — C_6H_5